BUILDING
METAL LOCATORS
A TREASURE HUNTER'S
PROJECT BOOK

No. 2706
$15.95

BUILDING
METAL LOCATORS
A TREASURE HUNTER'S
PROJECT BOOK

CHARLES D. RAKES

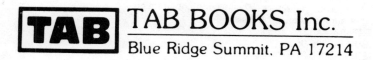

TAB BOOKS Inc.

Blue Ridge Summit. PA 17214

FIRST EDITION

FIRST PRINTING

Copyright © 1986 by TAB BOOKS Inc.

Printed in the United States of America

Library of Congress Cataloging in Publication Data

Rakes, Charles D.
    Building metal locators.

    Includes index.
    1. Metal detectors—Design and construction—
Amateurs' manuals. I. Title.
TK7882.M4R35 1986   622'.19   86-5859
ISBN 0-8306-0506-1
ISBN 0-8306-2706-5 (pbk.)

Contents

Introduction

It's only human nature to dream of riches and lost treasures, but most of us never have the opportunity to follow through on these universal dreams of striking it rich. Months, and even years, of research are necessary to search for a large or famous buried treasure. Even then who's to say that the treasure ever existed? Will it still be waiting just for you to discover?

Few real treasures that are found by the professional treasure hunter are reported to the public, and so many years of research and searching might go for naught. However, by arming yourself with a modern metal locator, taking a few hours of free time, and hunting in the right places, a good number of coins and other small valuables can be located. Any coins that are old enough to be made of silver will make even a small find a very good find indeed, since silver is worth several times its face value. Locating lost or buried coins might not constitute a large treasure find, but it can add up to a profitable outing for any ambitious metal-locator enthusiast.

The primary purpose of this book is to supply a number of tested and operating metal locator circuits that almost any electronics hobbyist, technician, or student can duplicate with comparative ease, and put to use in finding anything from valuable coins to large mineral deposits.

Chapter 1 introduces you to the different types of metal locators, and the basic operation of each type is explained. Chapters 2, 3, 4, 5, and 6 cover a variety of specialized metal locator circuits, including schematic diagrams, drawings, photographs, and helpful technical hints in construction and operation. Chapter 7 helps in selecting and using the proper type of metal locator for whatever purpose or job you have in mind.

Types of Metal Locators

There are two basic methods used to detect metal objects electronically. One method is to detect the change in inductance of a search loop when a metal object is moved into the loop's magnetic field; the other method is to detect the distortion in the magnetic field induced by the metal object.

BFO LOCATORS

The most common type of metal locator using the first method of detection is the BFO (*beat-frequency-oscillator*) *detector*. BFO metal locators are generally used for locating small metal objects at relatively shallow depths, but are not entirely limited to this type of use. Figure 1-1 illustrates the basic components of a BFO locator. The search loop oscillator and the reference oscillator are both operating near the same frequency, and their outputs are coupled to a common mixer circuit. The output of the mixer feeds an audio amplifier that drives a speaker, headset, or meter to indicate the presence of detected metal.

In operation, the search loop oscillator generates a slightly higher frequency than that of the reference oscillator by 50 to 100 Hz. The audio produces a low growling sound at the speaker, indicating a normal no-metal condition at the search loop. If the search loop is moved close to a ferrous metal object, the inductance of the search loop is increased slightly and the frequency of the loop oscillator goes down a few hertz. The audio sound at the speaker is

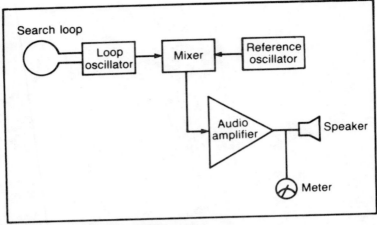

Fig. 1-1. Block diagram of BFO locator.

lower and indicates the presence of ferrous metal in the loop's field. The presence of a nonferrous metal object in the loop's field produces the opposite effect by reducing the inductance of the search loop. As the inductance is reduced, the frequency of the loop oscillator goes up, as does the audio signal produced at the output of the mixer, indicating the presence of nonferrous metal in the loop's field. The unique ability of the BFO locator to distinguish ferrous and nonferrous metal objects is very helpful indication when you are searching for valuable coins and other nonferrous objects.

BALANCED-BRIDGE LOCATORS

The *balanced-bridge metal locator* operates on the same principle as the BFO locator, but it interprets the inductance change by a different method. Figure 1-2 shows the basic components of a balanced-bridge metal locator. The search loop makes up one arm of a balanced inductance bridge and is fed a stable low-frequency signal from an internal oscillator. Without any metal in the loop's field, the bridge is balanced electronically so no tone is present at the speaker. When a metal object is moved into the loop's field, the bridge circuit is slightly unbalanced because of the small change in inductance of the loop. This unbalance produces a very small signal at the output of the bridge circuit. The signal is amplified many times and feeds the speaker and meter to indicate the presence of metal. Several types of specialized metal locators use this basic principle of operation.

The first method of metal detection discussed used the min-

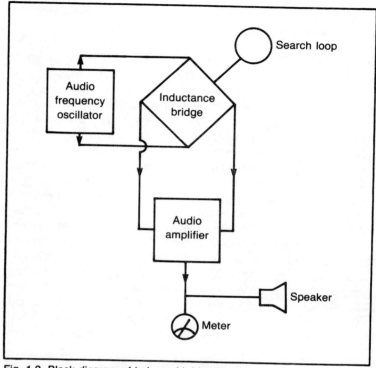

Fig. 1-2. Block diagram of balanced-bridge locator.

ute changes in the loop's inductance as an indication of metal present within the loop's field. The frequency-shift method of detection, or the more common BFO locators, and the balanced bridge locators are the two basic types of circuits that make use of these operating principles. Chapters 2 and 3 will offer several locators of this type to build, and will give a circuit-by-circuit explanation of each metal locator.

TRANSMITTER/RECEIVER LOCATORS

The second family of metal locators operates on the field distortion principle and can be separated into two circuit categories. The first type of locator is the simplest of the two, and it began its career back in 1929 when Gerhard R. Fisher received a patent on an electronic device for locating buried metal objects. This was the first *transmitter/receiver metal locator* and the first successful electronic metal locator to be produced commercially.

3

Figure 1-3 is a block diagram of a simple transmitter/receiver metal locator. The transmitter and receiver are separate and independent circuits and are usually separated by 3 to 4 feet. The receiver is positioned in a horizontal plane and the transmitter in a vertical position. An important characteristic of the loop antenna coil is its directional properties, with the maximum radiation in an edgewise direction. Almost no energy is radiated in the direction perpendicular to the loop. The receiver's lop has the same directional characteristics as the transmitter loop, and receives little or no energy from the transmitter's magnetic field. Figure 1-4 illustrates the magnetic field radiated by the locator's transmitter loop. As shown in the illustration, the majority of the rf energy is radiated circular in a donut fashion around the transmitter's loop, and almost no energy is directed toward the receiver's loop. The dotted lines in Fig. 1-4 show the sensitive area of the receiver's loop.

A metal object positioned in either the transmitter or receiver loop's field will cause the magnetic field to become slightly distorted, and the receiver will pick up a portion of this redirected energy and give a positive indication of metal. T/R locators are generally suitable for deep penetration and do not readily respond to small metal objects. Their main use is in locating and tracing underground pipes and for finding large mineral deposits. Large

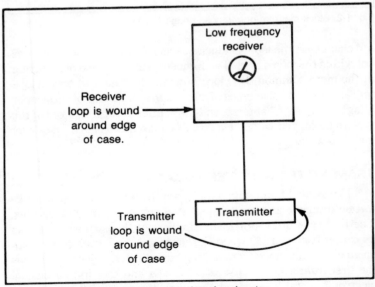

Fig. 1-3. Block diagram of transmitter/receiver locator.

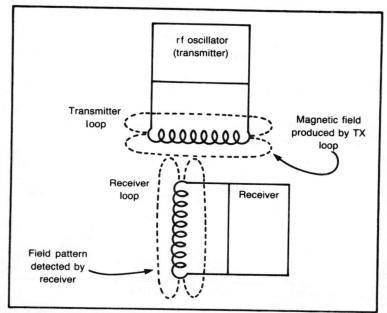

Fig. 1-4. Magnetic field of T/R locator.

treasure troves can be located with a T/R locator if the size and depth are within the detection range of the locator that you use.

INDUCTION-BALANCE LOCATORS

The second type of metal locator using the distorted field principle is the *induction-balance circuit*. A block diagram is shown in Fig. 1-5. Three individual loops are used in this circuit. Two of these loops are fed out of phase with a common signal from a rf oscillator. The third, or middle, loop is connected to a high-gain rf amplifier, receiver, and detector circuit. The amplified rf signal is detected and drives an audio amplifier and meter circuit to indicate the presence of metal. The two transmitter loops are mechanically and electrically balanced so little or no rf is present at the receiver's loop.

The three loops are stacked one above the other, and when a metal object is positioned below the search loop, a slight unbalance occurs in the bottom transmitter's loop. The unbalanced condition induces a small rf signal in the receiver's loop. This signal is amplified and detected, indicating a metal object is in the search loop's field.

5

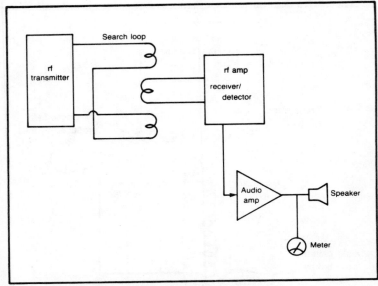

Fig. 1-5. Block diagram of induction-balanced locator.

INDUCTION-BALANCE
LOCATORS WITH COPLANAR LOOP

Figure 1-6 shows another locator that operates on the *VLF* (very low frequency) induction balance principle. This locator uses a specially designed search loop. The *coplanar loop* is designed to eliminate one of the transmitter loops and to simplify the overall circuit design. This loop is formed by taking a single transmitter loop and folding back a section of the loop to create a balanced magnetic field area for the receiver's loop to be placed in. The circuit operation of the coplanar locator is very similar to the locator in Fig.1-5, but operates at a much lower frequency. The VLF mode of operation helps to overcome some of the major problems found in earlier locator circuits. The problem of ground effect is greatly reduced at frequencies below 10 kHz, and the ability to discriminate between good metal and trash is enhanced many times at these very low frequencies.

Chapters 4 and 5 offer several different T/R and VLF induction balance metal locators that can be built to meet almost any searcher's requirements. Chapter 6 consists of several unusual "maverick" locators that do not fit comfortably in any of the other categories, but are too good not to pass on to the experimenter and serious builder.

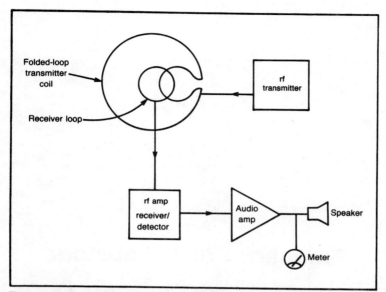

Fig. 1-6. Block diagram of induction-balanced locator with coplanar loop.

Frequency Shift Locators

The BFO electronic metal locator is by far the most popular of the frequency shift locators and one of the simplest to build and use successfully. The first BFO construction project described is designed to meet the following goals:

- Good to excellent sensitivity to small objects, such as coins and jewelry.
- Stable operation.
- Simple to operate.
- Minimum ground effect.
- Simple and economical to build and operate.

These are fairly modest requirements, but even so, several compromises must be made to even come close to achieving them.

Selecting the size of the search loop depends on what type of metal objects the locator is designed to search for, and their size and depth. As the diameter of the search loop decreases, the sensitivity to small metal objects increases, but the depth of penetration also decreases. A good compromise is the 9-inch-diameter loop selected here.

There are several advantages to using a single-turn search loop coil. The single-turn loop design almost eliminates the ground effect, and without the normal loss in sensitivity of multi-turn loops. In most BFO locators, the loop is made up of several turns of copper wire, and is surrounded by a Faraday shield to reduce the ground effect. When a coil or loop is covered with a Faraday shield, the overall sensitivity and penetration is degraded to some extent because of the eddy currents induced by the metallic Faraday shield. However, this problem is offset by a significant reduction in ground effect.

Ground effect occurs when the search loop is moved toward the ground to begin searching. When the loop approaches the surface of the earth, the capacity between the loop and earth tends to change the frequency of the loop's oscillator, as if there was an actual change in internal capacitance.

The single-turn loop metal locator overcomes this problem by using a very large "C" (capacitor) to "L" (inductance) ratio in the search oscillator's circuit. The actual capacitance change when the loop is moved toward the ground is very small. With the large capacitor used to tune the loop's oscillator circuit, very little or no frequency change occurs from the ground effect. Additional circuit stability is obtained by using a high-impedance Darlington transistor amplifier circuit to isolate the low-impedance tuned-loop circuit from the influence of the active components of the oscillator circuit. Also, by locating the loop oscillator circuit at the base of the single-turn loop, overall stability will be increased greatly and the BFO circuit will be able to operate in its most sensitive and stable mode.

CONSTRUCTION OF THE SINGLE-TURN BFO LOCATOR

Figure 2-1 shows the single-turn BFO metal locator. The actual construction need not be as shown, but you should follow the precautions and guidelines offered here if you do decide to use a different approach. Start construction by fabricating the search loop. Figure 2-2 shows the round, pressed-wood form that holds the loop and the loop's oscillator circuit components. Almost any workable nonmetallic material will do for the loop's form. Cut or saw a 9-inch circle from the 1/2-inch material and cut a groove in the outside edge of the form to allow the copper loop to fit firmly in place. Take a 27 1/2-inch length of 1/4-inch copper tubing and form a 9-inch diameter loop by using the grooved form as a bending guide. This should leave a gap between the loop's ends of about 1 inch. If cutting a groove in the outside edge of the form proves to be a problem, then make the form slightly larger and mount the loop on top; either mounting method will work just fine.

Fig. 2-1. BFO locator with single-turn loop.

If for some reason your loop departs somewhat from dimensions shown, the frequency of the loop oscillator could be beyond the tuning range of the reference oscillator and the locator would fail to operate. If the loop comes out larger in diameter, then the search oscillator will be operating at a lower frequency. To bring the oscillator back within the range of the reference oscillator reduce the values of capacitors C1 and C2 by a small amount. Ten to twenty percent is a good starting point. If the loop is too small, then just increase the values of C1 and C2 until you obtain the desired frequency. Actually, you should not need to change the capacitors if the loop diameter varies by no more than ± 1 inch, as the tuning range of the reference oscillator is great enough to allow for minor variations in the loop dimensions.

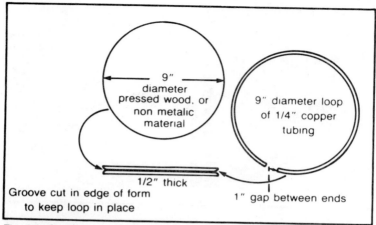

Fig. 2-2. Construction of single-turn loop.

If a frequency counter is available the search loop diameter can be changed to any desired size and be retuned by selecting the values of C1 and C2 to bring the oscillator within the tuning range of the reference oscillator. The search loop can be made very much larger for deep searching, or much smaller to locate a very small object near the earth's surface. In either case a frequency counter and oscilloscope will be helpful in experimenting with different loop sizes and circuit variations.

If the groove method of holding the loop in place is used, the ends of the copper tubing can be kept in place by lacing a few turns of bare copper wire through two holes drilled in the form (see Fig. 2-3), and then soldering to hold it rigidly in place. At the same time

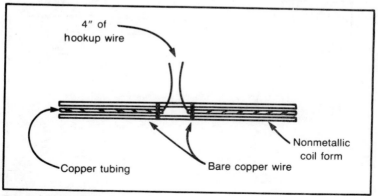

Fig. 2-3. Detail of loop construction.

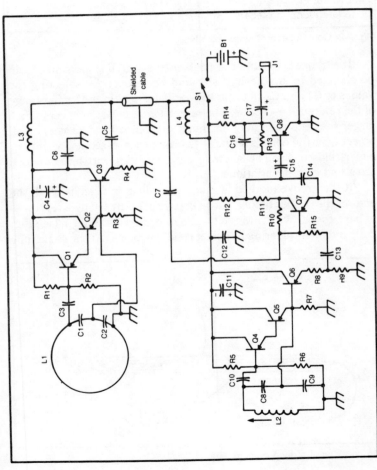

Fig. 2-4. Schematic diagram of single-turn BFO locator.

solder a 4-inch length of hookup wire to each end of the loop to go through the plastic case and connect to the oscillator circuit.

The perf-board and bottom section of the plastic case is mounted to the loop's form with two 3/4-inch 6-32 screws. Drill and tap two holes in the loop's form to match up with the perf-board and plastic case. Figure 2-4 is the complete circuit diagram for the BFO locator, and Table 2-1 is the parts list. Proceed to the drawing in Fig. 2-5 for parts placement for the loop oscillator circuit. Here a neat wiring job is a must.

Mount the search loop components on the perf-board with flea clips. Keep all component leads short and all wiring point to point and mechanically rigid. Any movement here will cause the search oscillator to be unstable and will keep the locator from operating at its best sensitivity and stability. The remaining circuit components are housed in a Radio Shack 4- × -2 3/8- × -6-inch metal cabinet, and are mounted on a 3 1/2- × -4-inch section of perf-board. Figures 2-6 and 2-7 show a suggested parts layout. The loopstick, phone jack, and power switch are mounted on the front of the cabi-

Table 2-1. Parts List for Single Turn BFO Locator.

B1	9-volt battery (6 AA cells in series)
C1, C2, C12	.27 µF/100-volt mylar or similar capacitor
C3, C5, C6	.1 µF/100-volt mylar or similar capacitor
C4, C17	22 µF/16-volt electrolytic capacitor
C7, C13, C16	.0015 µF/100-volt mylar or similar capacitor
C8	330 pF/50-volt polystyrene capacitor
C9	660 pF/50-volt polystyrene capacitor
C10	.01 µF/100-volt mylar or similar capacitor
C11	47 µF/16-volt electrolytic capacitor
C14	.0036 µF/100-volt mylar or similar capacitor
C15	10 µF/16-volt electrolytic capacitor
Q1-Q8	2N3638 pnp silicon transistor
L1	Search loop (see text for materials)
L2	Broadcast band loopstick (adjustable)
L3, L4	4.7 mH choke coil
J1	1/4-inch phone jack
R1, R2, R5, R6, R13	220 kilohm 1/4-watt 5% carbon resistor
R3	470 ohm 1/4-watt 5% carbon resistor
R4, R7, R14	2.2 kilohm 1/4-watt 5% carbon resistor
R8, R11, R15	1 kilohm 1/4-watt 5% carbon resistor
R9	270 ohm 1/4-watt 5% carbon resistor
R10	100 kilohm 1/4-watt 5% carbon resistor
R12	4.7 kilohm 1/4-watt 5% carbon resistor
S1	SPST toggle switch
Miscellaneous	2 kilohm headphones, cabinets, loop form, etc. (see text for additional parts details)

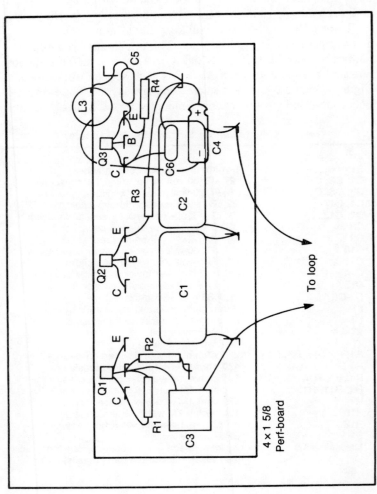

Fig. 2-5. Parts placement for loop oscillator circuit.

14

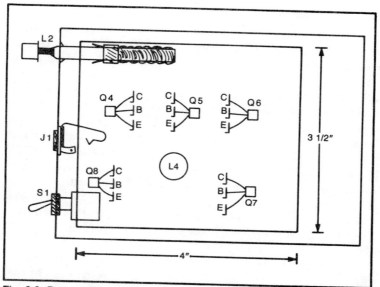

Fig. 2-6. Parts placement on main perf-board.

net. The metal cabinet can be mounted to the plastic handle, or as shown equipped with a clip for carrying on your belt. The same wiring practices apply for the circuit components in the main housing, as any parts movement here will also cause unstable operation. Each component part and all circuit wiring must be rigid and should be double-checked against the schematic diagram before applying power. A section of RG174/U coaxial cable connects between the search loop oscillator circuit and the circuitry in the main cabinet.

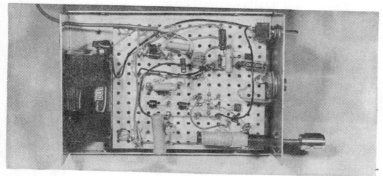

Fig. 2-7. Parts placement in main cabinet.

15

Circuit Operation

To help in better understanding and troubleshooting the BFO locator, here is a brief description of the circuit operation. Starting with the loop and search oscillator circuit, L1, C1, and C2 make up the tuned circuit for the single-turn search loop. Transistors Q1, Q2, and Q3 are in a high input impedance Darlington amplifier configuration to isolate the tuned circuit from the active devices. Resistors R1 and R2 set the bias for the three transistors. The rf signal is coupled from the emitter of Q3 through C5, and is fed down the shielded cable to the mixer circuit located in the metal cabinet. L3 is a 4.7 mh choke and provides rf isolation. It allows the battery voltage to be fed through the same shielded cable without interfering with the rf signal. Capacitors C7 and choke L4 isolate the mixer circuit. Capacitors C4 and C6 are for additional rf and battery filtering. The loop oscillator operates at approximately 600 kHz.

The reference oscillator is very similar to the search loop oscillator circuit, and its frequency of operation is determined by the values of L2, C8, and C9. The frequency of the reference oscillator can easily be varied over a 100 kHz range with the slug adjustment of L2 to match the frequency of the search loop oscillator. Transistors Q4, Q5, and Q6 isolate the tuned reference circuit.

The impedance of the reference oscillator tuned circuit is much greater than that of the search loop circuit, and a larger rf signal is developed at the emitter of Q6. Resistors R8 and R9 make up a voltage divider that supplies the mixer circuit an rf signal level approximately the same as that received from the search oscillator. The rf signal from the reference oscillator is coupled to the base of the mixer transistor Q7 through C13 and R15. The rf signal from the loop oscillator is coupled through C7 to the mixer's base. The nonlinear mixing action offered by the base emitter junction of Q7 produces a sum and difference frequency at the collector. Only the low frequency component is passed through to the audio amplifier stage, Q8. Resistors R11, R12, and C14 make up a low pass filter, cutting off the frequencies above the audible range and passing only the low audible tones to the input of the audio amplifier. The audio output is coupled through C17 to the headphone jack. Any quality high-impedance headphones will work fine with the locator.

Checking Out the Locator

Double-check all electrical and mechanical connections and solder joints; be sure each transistor and diode is connected correctly, and all wiring matches the schematic diagram.

Place the batteries in the battery holder and connect to the circuit, plug in headphones, and set L2's slug to approximately half way in the coil. With everything hooked up and the loop positioned away from any metal object, turn the power switch on and slowly adjust the tuning slug in or out. You should hear several beat notes while tuning, but continue to tune L2 until you have located the loudest of these tones. If all circuits are working properly the tuning range of L2 will be within the frequency range of the loop oscillator, but if a loud beat tone is not heard, or only low volume tones are heard, try to determine the frequency of the two oscillators. A frequency counter would be the best method to use in checking the frequency of each oscillator, but a standard AM broadcast receiver can be used to check the second harmonic of each oscillator. If a frequency counter is available connect the input to the emitter of Q3 to check the loop oscillator and to the emitter of Q6 to check the reference oscillator.

The frequency of either oscillator can be changed to bring it within the range of the other, but the simplest and best method to use is to lower the oscillator that is operating at the highest frequency down to the same frequency as the other oscillator.

If the search loop oscillator is operating at the higher frequency of the oscillators, add capacitance to both C1 and C2 in equal amounts until the desired frequency is obtained. Before changing the frequency of the reference oscillator reset L2's slug to mid position, then add capacitance to C8 and C9 in equal values until the desired frequency is obtained. If the trouble is not due to the frequency setting, refer to the voltage readings in Table 2-2. All voltage readings are taken with a 20 kilohm-per-volt meter. The positive meter lead is connected to circuit common or battery positive. All voltages are taken with negative probe, with the locator power on and the loop away from any metal items. All readings would be

Table 2-2. Voltage Readings for Single Turn BFO Locator.

Circuit Location	Voltage
Emitter of Q2	− 3.4 volts
Emitter of Q3	− 2.8 volts
Emitter of Q5	− 4.0 volts
Emitter of Q6	− 3.5 volts
Collector of Q7	− 2.6 volts
Collector of Q8	− 5.0 volts

within ±20% of the values given here.

Take the locator outdoors and position the search loop about two inches above the ground, and adjust L2 until a low-frequency beat note is heard. If everything is stable and the circuit is operating as it should, you can adjust the reference oscillator to within a few cycles of the search oscillator, and will hear a low frequency "put-put" in the phones. This is the most sensitive mode of operation for the BFO locator. Practice with various coins and small metal objects to familiarize yourself with the locator's operation. It takes a fair amount of time to become proficient with any metal locator, so you will need to practice if this is your first BFO metal locator.

LARGE LOOP BFO LOCATOR

The second metal locator is shown in Fig. 2-8. This frequency-shift metal locator is designed to search for large metal objects or

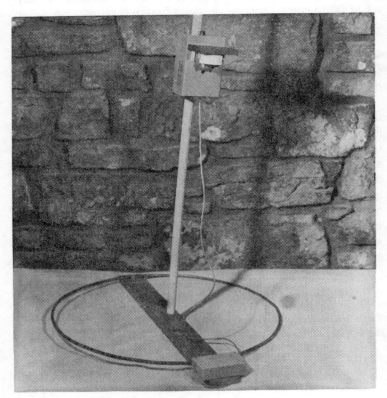

Fig. 2-8. BFO locator with large loop.

mineral deposits at depths of several feet. The search loop is 24 inches in diameter and is of the single-turn type. Single coins and other small objects will be overlooked by the deep searching 24-inch loop. This feature is a real plus to the treasure hunter who is looking for a large buried object, and doesn't want to bother digging up small metal items.

The schematic diagram of the large loop BFO locator is shown in Fig. 2-9 and Table 2-3 is the parts list. The search loop oscillator, reference oscillator, and the mixer circuit all operate like their counterpart circuits in the first BFO locator, but the remaining circuit operation is unusual and completely different. Transistors Q7, Q8, and the associated components form a twin-T notch filter that sucks out a narrow band of frequencies near 700 Hz. All other audible signals coming from the mixer are passed on through the twin-T filter to the meter and headphone circuits. With the two oscillators set apart in frequency by 700 Hz the audio signals reaching the meter and headphones are at a minimum. Under these settings the meter will indicate a near zero reading. The bandwidth of the twin-T filter is only a few Hz wide and any small shift in the loop oscillator's frequency will result in a large change in the meter reading.

This circuit feature makes the large BFO a very sensitive metal locator. A very large metal object, like an automobile, above ground will give an indication at a distance of over six feet. Underground detection of a similar size metal object, will in many cases, show up even better than above ground.

Start by forming a 75-inch section of 5/16-inch copper tubing into a 24-inch circle, with an end gap of 2-inches. You can easily do this if a bucket, barrel, or other round object is used to roll the tubing around to form a circle. Figures 2-10 and 2-11 will be helpful in building the loop. After the loop has been formed it will try to spring out to a larger size. A simple method to use in holding the loop to size, is to run a small wire through the tube and tie a small loop in one end of the wire. Run the other end of the wire through the small wire loop and pull on the wire until the search loop ends are about two inches apart. Tie the ends of the wire together and the loop will stay in position until it is mounted in place on the support board. After the loop is mounted in place cut and remove the wire.

The remaining construction of the loop oscillator circuit is shown in Fig. 2-12. The part layout doesn't have to match the one shown, but good mechanical and electrical practices must be fol-

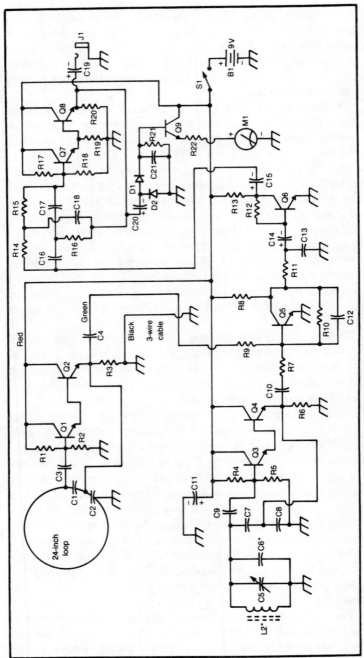

Fig. 2-9. Schematic diagram of BFO locator with large loop.

Table 2-3. Parts List for Large Loop Locator.

B1	9-volt transistor battery
C1, C2	.12 µF/100-volt mylar capacitor
C3, C13	.047 µF/100-volt mylar capacitor
C4, C10, C12	680 pF/100-volt disc. ceramic capacitor
C5	365 pF variable capacitor (broadcast type)
C6	Optional additional capacitors (see text)
C7	.068 µF/100-volt mylar capacitor
C8	.039 µF/100-volt mylar capacitor
C9	.015 µF/100-volt mylar capacitor
C11	330 µF/16-volt electrolytic capacitor
C14	4.7 µF/16-volt electrolytic capacitor
C15, C20	6.8 µF/16-volt electrolytic capacitor
C16, C17	.022 µF/100-volt mylar capacitor
C18	.044 µF/100-volt mylar capacitor
C19	10 µF/16-volt electrolytic capacitor
C21	.1 µF/100-volt mylar capacitor
D1, D2	1N914 silicon diode
Q1-Q9	2N5249 npn silicon high-gain signal transistor, sub 2N5088
L1, L2	Handmade coils (see text for details)
M1	0-200 µA dc meter
J1	1/4-inch phone jack
R1, R2, R4, R5	220 kilohm 1/4-watt 5% resistor
R3, R6	470 ohm 1/4-watt 5% resistor
R7, R9, R11	1 kilohm 1/4-watt 5% resistor
R8, R13, R16, R22	4.7 kilohm 1/4-watt 5% resistor
R10, R12, R17, R18	470 kilohm 1/4-watt 5% resistor
R14, R15, R19	10 kilohm 1/4-watt 5% resistor
R20	2.2 kilohm 1/4-watt 5% resistor
R21	100 kilohm 1/4-watt 5% resistor
S1	SPST toggle switch
Miscellaneous	2 kilohm headphones, cabinets, copper tubing, hardware, etc. (see text for additional data)

lowed to assure a stable, operating locator.

The reference oscillator coil is hand wound and when completed should be similar to the one illustrated in Fig. 2-13. Take a 1 1/2-inch section of 1/4-inch ferrite rod (loopstick material will do) and wind 10 turns of #18 plastic-covered, single-strand copper wire around the rod. Keep the turns tight and close together to form a mechanically stable coil. Take the two ends of the wire (plastic cover in place) and twist together about two or three turns. This will keep the coil from coming unwound and will not hurt performance. Leave about 3/4-inch of wire at each end of the coil for connecting to the circuit perf-board pins. If for some reason your coil doesn't have the correct value of inductance the reference oscillator can be tuned to the desired frequency by adding or removing capacitance in the same manner as with the first BFO locator.

The circuit components located in the metal cabinet can follow the suggested layout shown in Fig. 2-14. Only the major parts are shown in this drawing to help in the basic layout of the component parts. The completed circuit is shown in the photo in Fig. 2-15.

Capacitors C16, C17, and C18 should be mylar in type and have a value tolerance of ±5% or better. Do not change any of the capacitor or resistor values in the twin-T filter circuit, because the components used must be balanced and matched in value to produce a sharp and narrow filter response.

Any dc microammeter can be used for M1. If a 200 μA meter is available a 4.7 kilohm resistor will be needed for R22. A 1 mA meter requires a 1 kilohm resistor for R22. The actual resistor value is not critical, as long as the meter is not damaged when it is pegged

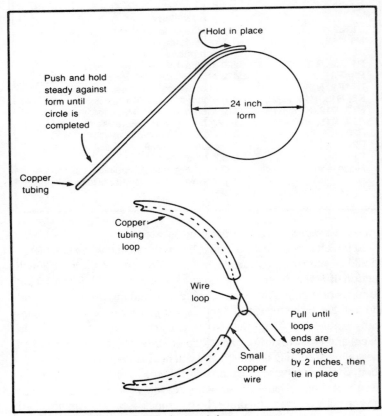

Fig. 2-10. Construction of 24-inch search loop.

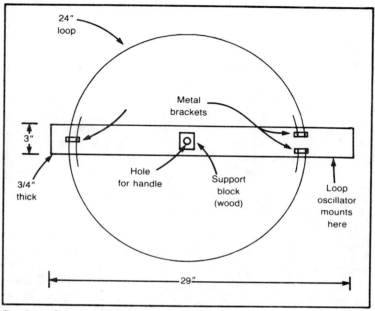

Fig. 2-11. Support of large loop.

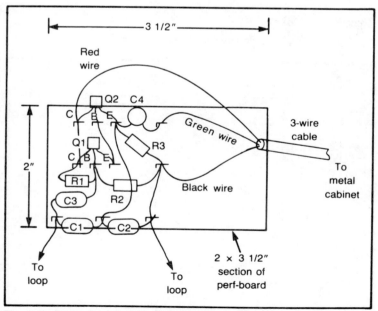

Fig. 2-12. Parts placement for large loop locator.

23

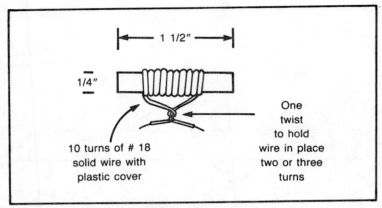

Fig. 2-13. Building the reference oscillator coil (L2).

over scale. Leave room for additional capacitors (indicated in the schematic by C6) across L2, as the actual values needed can vary from the ones shown. The frequency of the loop oscillator and reference oscillator will be somewhere around 550 to 600 kHz. The actual frequency is not important as long as both oscillators can be tuned to the same frequency.

Circuit Operation

The following basic theory of operation will be helpful in get-

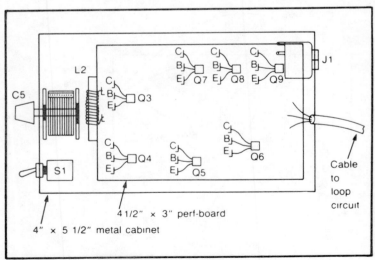

Fig. 2-14. Parts placement for main perf-board.

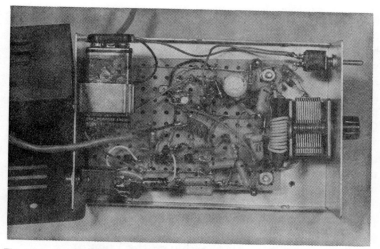

Fig. 2-15. Parts placement in main cabinet.

ting your locator working. L1, C1, and C2 form a very low impedance tuned circuit for the loop oscillator. Transistors Q1 and Q2 offer a very high input impedance to the tuned circuit, and produces very little loading on the loop's tuned circuit. This combination of a low-impedance tuned circuit and a high-impedance transistor oscillator and amplifier circuit produces an extremely stable oscillator that is ideal for metal locator use. The reference and search loop oscillator circuits are very similar in design, and any temperature or voltage variations will have the same effect on each of the oscillators. The frequency shift of each oscillator will be in the same direction and the difference in frequency at the output of the mixer will remain unchanged. The errors induced by both oscillator circuits are self-cancelling under these circuit conditions.

The rf signal for the mixer is sampled at the emitter of Q2 and is fed to the main circuit through the three-wire cable that also carries the battery voltage to the search loop circuit. L2, C7, C8, C5, and C6 are the tuned circuit components used in the reference oscillator, and the signal for the mixer is taken at the emitter of Q4.

Transistor Q5 mixes the two rf signals. The high-frequency content of the mixer's output is fed back to the input through C12 to reduce the high-frequency response of the mixer and to allow only the low frequency audio to pass. R11 and C13 provide additional low-pass filtering. Q6 increases the audio level sufficiently to drive the filter circuit and headphones.

R14, R15, R16, C16, C17, and C18 form a twin-T notch filter circuit set at 700 Hz. Q7 and Q8 provide a high-input-impedance buffer amplifier for the filter circuit to keep it's Q high and bandwidth narrow. All audio signals, with the exception of a narrow band of frequencies around 700 Hz, are fed to the headphones and to a voltage-doubler circuit that drives the meter circuit. The dc output from the voltage doubler is buffered by an emitter follower, Q9, to drive the meter without loading the twin-T filter circuit.

Tuning the reference oscillator to produce a difference frequency of 700 Hz at the output of the mixer will cause the output of the twin-T filter to drop to a very low output level. The meter will respond in a similar manner and drop to a low scale reading. Any frequency shift in the loop oscillator will cause a large change in the meter reading and headphone output.

Final Positioning and Assembly of the Locator

After the wiring is completed and checked against the diagram in Fig. 2-9, connect the battery and plug in the headphones. Turn the power on and move the search loop clear of any metal objects. Tune the 365 pF variable capacitor (C5) until an audible tone is heard in the headphones. Actually, you will be lucky if both tuned circuits are within 3000 Hz of each other, as this is the maximum tuning range of C5. The search loop oscillator's frequency should be close to 600 kHz and it would be best to leave it alone and re-tune the reference oscillator to the same frequency as the loop oscillator. If the reference oscillator is operating at a higher frequency than the search loop oscillator, start adding capacitors (C6) of 330 pF, one at a time until the variable capacitor (C5) tunes to the desired frequency. If for some reason the frequency of the reference oscillator is too low, change C7 to a smaller capacitor value. If you start out with a .056 mfd capacitor in place of C7 this should place the reference oscillator above the frequency of the search oscillator. If so, just follow the previous suggestion of adding capacitance to bring the frequency of oscillation into the tuning range of C5. A frequency counter would make the job of tuning much easier. If any problems other than operating frequencies occur, refer to the voltages given in Table 2-4. All voltages are taken with a 20 kilohm-per-volt meter. The negative lead of the meter connects to the negative battery or ground side of the circuit. Voltages are taken with power on and search loop away from all metal objects. Measurements should be within ± 20% of the following readings.

Table 2-4. Voltage Readings for Large Loop Locator.

Circuit Location	Voltage
Emitter of Q2	3-volts
Emitter of Q4	3-volts
Emitter of Q8	2-volts
Collector of Q5	3.5-volts
Collector of Q6	4.8-volts

Using the Deep-Seeker Locator

The operation of the deep searching locator is similar to that of the smaller BFO locator, with the exception of the meter and selective filter circuit. Using the meter and filter circuit will let you take full advantage of the locator's super-sensitive mode of operation.

Tune the reference oscillator with C5 until the meter moves toward a full scale reading, and continue tuning in the same direction until the meter's needle takes a fast swing toward zero. Tuning C5 in either direction should bring the meter's needle back up scale rapidly. The two oscillators are now operating with a frequency difference of 700 Hz. The most sensitive operation occurs about 1/5 of the way up the scale from the dip point.

Capacitor C5 can be tuned so that nonferrous metal will cause the meter to read up scale, and ferrous metal down scale, or vice versa. Once these directions have been determined for ferrous/nonferrous metals, mark the setting on the face of the locator and use it to determine the type of metal that is detected. The strongest signal, or greatest change in the meter reading, will occur when the loop is positioned right over the center of the buried object.

CRYSTAL FILTER LOCATOR

The third frequency shift metal locator uses only a single oscillator and requires neither a reference oscillator or a mixer circuit to function. The sensitivity of the crystal filter locator (CFL) is better than most BFO circuits and uses fewer parts. A super-selective filter circuit is used to detect any frequency shift of the search oscillator; the amount of shift is indicated on a dc meter.

Theory of Operation

Figures 2-16 and 2-17 illustrate the simplicity of the crystal filter

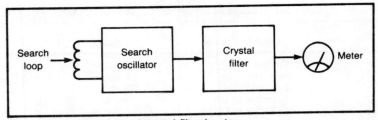

Fig. 2-16. Block diagram of crystal filter locator.

locator circuit. The parts list is shown in Table 2-5. Transistor Q1, search loop, and associated components form a Colpitts oscillator circuit. The frequency of operation is determined by the values of C1, C2, C3, and the inductance of the search loop. The output of the search loop oscillator is coupled to an emitter-follower amplifier, Q2, that isolates the filter circuit from influencing the frequency of the oscillator by pulling when both are on the same resonant frequency. If the coupling between the oscillator and crystal becomes too great the crystal will actually lock the oscillator to the crystal's frequency. The locator can operate in this mode but sensitivity will be greatly reduced. The output of the crystal is rectified by the voltage doubler of circuit D1 and D2 and is fed to the meter circuit through an emitter follower, Q3. The maximum meter current is set by R4, and will vary with the type of sensitivity of the meter used.

With the oscillator operating at 1.25 MHz (crystal frequency used) the maximum rf signal will pass through the crystal. The rf

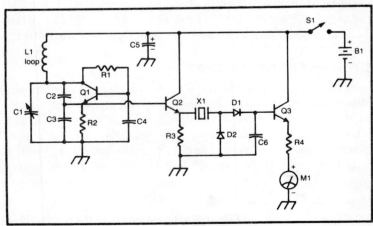

Fig. 2-17. Schematic diagram of crystal filter locator.

Table 2-5. Parts List for CFL Locator.

B1	9-volt transistor battery
C1	50 pF to a 365 pF tuning capacitor (see text)
C2, C3	.0036 μF/100-volt mylar capacitor
C4	.015 μF/100-volt mylar capacitor
C5	22 μF/16-volt electrolytic capacitor
C6	.1 μF/100-volt mylar capacitor
Q1, Q2, Q3	2N5249 or 2N5088 npn silicon transistor
D1, D2	1N914 silicon diode
M1	0-500 μA dc meter
R1	220 kilohm 1/4-watt 5% resistor
R2, R3, R4	1 kilohm 1/4-watt 5% resistor
XTL	1.25 MHz transmitting crystal (see text)
S1	SPST toggle switch
Miscellaneous	Cabinet, perf-board, hardware, etc. (see text for misc. parts info.)

signal will be changed to dc by the voltage doubler and a full-scale reading will appear on M1. Assume that no metal is near the loop and the oscillator is tuned to the crystal resonance frequency, an the meter reads full scale. If the search loop is moved over a metal object, the search oscillator will shift in frequency, the rf passing through the crystal will be reduced and the meter will accordingly read lower. The search oscillator can be tuned to the high or low frequency curve of the crystal's bandpass and be used to distinguish between ferrous and nonferrous metals.

Building the CFL Metal Locator

Figure 2-18 shows the complete locator. A 5 × 3 × 2-inch deep drawn aluminum case houses all parts but the search loop. Any comparably sized aluminum cabinet will serve to house your circuit. The search loop is 5 inches in diameter and consists of five turns of number 24 enameled copper wire, and is wound on a needle point wood form. The loop is supported by a wooden form and plastic pipe. It can be constructed by following Fig. 2-19.

The components are mounted on a 4 × 1-5/8 inch section of perf-board on the inside of the metal cabinet (see photo in Fig. 2-20). The actual component layout isn't critical—just keep the component leads short and follow the general layout in Fig. 2-21. Capacitor C1 can be any value from 50 pF to a standard 365 pF broadcast tuning type capacitor. If a value larger than 100 pF is used you might want to place a 180 pF in series with the variable capacitor (C1) and the loop. Doing this will reduce the overall value of C1 and will give a better fine-tuning action, and will make setting the oscillator to the desired frequency much easier. If C1 is too large the tuning of the oscillator will become tedious and will not operate at the most sensitive setting.

After the wiring and mechanical work is finished, place a battery in the holder, turn the power on, and with the loop away from any metal object, tune C1 and watch for a quick rise of the meter's needle. If the oscillator frequency happens to be out of range of the crystal frequency, retune by changing the values of C2 and C3. A frequency counter would be helpful in selecting the values for C2 and C3. If a counter isn't available use an AM receiver and tune to a station near the crystal frequency, then retune the oscillator until you hear a beat tone on the radio.

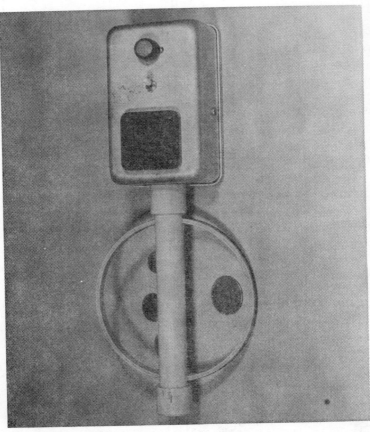

Fig. 2-18. Crystal filter locator.

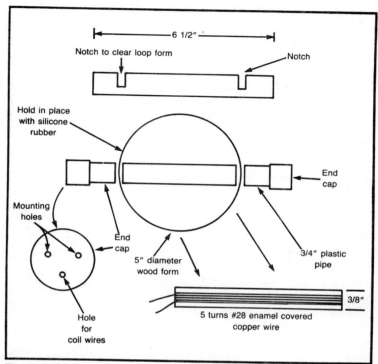

Fig. 2-19. CFL search loop.

Fig. 2-20. Inside the crystal filter locator.

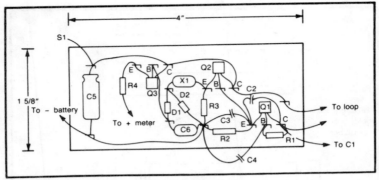

Fig. 2-21. Parts placement for crystal filter locator.

The AM radio can also be used to determine if the oscillator frequency is too high or too low by tuning the radio to the carrier produced by the loop oscillator. If the oscillator happens to be on the same frequency as a standard AM station you will hear a beat note. If it happens to be on a clear channel you won't hear anything. If the frequency is too high, add a small value of capacitance to C2 and C3 in equal amounts. If the frequency is too low, lower the values of C2 and C3 about 20 to 30% and retune to bring the oscillator into range. You can add or remove small values of capacitors in small steps to obtain the desired frequency. The actual frequency of operation isn't very critical as long as the oscillator will tune to the crystal's resonant frequency.

The 5-inch loop is suitable for locating small or medium-sized objects at relatively shallow depths. An ideal use of the small locator would be in searching for hidden objects in the walls of old houses or out-buildings. Another use could be in searching for concealed weapons, or as a companion to a larger locator.

If you have the means of determining and setting the frequency of the search oscillator, the loop can be changed to any size. The main circuit requirement for the CFL locator is that the loop oscillator be able to tune to the crystal's resonant frequency. Almost any crystal with a series-resonant frequency of 25 kHz to 3 MHz will work. A coin shooter locator can be changed into a deep searcher by increasing the loop's size, or the loop can be reduced to under one inch for seeking out very small objects. With the aid of a frequency counter or an AM radio to help in tuning the loop's oscillator, the locator can be modified to fill almost any special requirement.

The next two frequency shift metal locators are really designed for the serious electronic hobbyist. The first circuit is an ultra-simple and very sensitive single-transistor CFL locator. The one special component that makes the circuit possible is the sensitive 50 μA dc meter. The search loop is 24 inches in diameter, and has 6 turns of number 20 enameled copper wire. A special Faraday shield is wound around the outside of the loop. The loop's form is a wood hoop that can be found in almost any general hobby shop. This simple metal locator will rival the best BFO circuits in sensitivity and will outperform many of the high-cost locators on the market. The search loop is designed for deep searching, but by building a smaller loop, the circuit will also work well as a coin-shooter locator. The frequency of operation need not be exactly 200 kHz as used in this locator, but should stay within a range of 100 kHz to 300 kHz and match the crystal to be used.

ONE TRANSISTOR LOCATOR

A few construction hints should be all that are needed to build your own version of this simple locator. Start by making the search loop first. Take the 24-inch wood form and wind six turns of number 20 enameled copper wire in a side-by-side fashion around the form; don't "jumble wind" the coil. After the coil is completed, cover with one layer of black plastic electrical tape. See Fig. 2-22 to aid in the loop construction.

The Faraday shield is made by winding about 68 turns of number 20 bare copper wire, evenly spaced, around the search loop. Take another piece of bare copper wire and solder it to each turn on the inside of the Faraday shield coil's form. Be sure that a 2- or 3-inch gap is left between the ends of the shield. Connect one end of the Faraday shield to one end of the coil, and this common connection to the positive battery side of the circuit as in the diagram.

The circuit components should be mounted on a section of perf-board of sufficient size and mounted in a metal cabinet large enough to house the meter, perf-board, battery, and other parts without crowding. The most important circuit requirement is for the loop oscillator to tune to the crystal's resonant frequency. If the oscillator will tune to the resonant frequency the locator should perform without any difficulties; if not go over the simple circuit and see if an error has been made.

The simplest BFO metal locator requires only two oscillators,

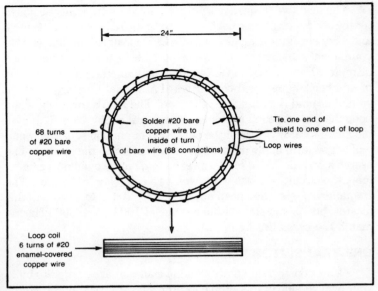

Fig. 2-22. Loop construction for single transistor locator.

a mixer, and a sensitive set of headphones. The complete parts list appears in Table 2-6, and the circuit shown in Fig. 2-23 is just a simple BFO locator circuit, and is quite suitable for coin-shooting.

Table 2-6. Parts List for Single Transistor Locator.

B1	9-volt transistor battery
C1, C2	.01 μF/100-volt mylar capacitor
C3	365 pF tuning capacitor (see text)
C4, C6	.1 μF/100-volt capacitor, mylar type
C5	47 μF/16-volt electrolytic capacitor
D1, D2	1N914 silicon diode
Q1	2N5249, 2N5088, or 2N2222 npn transistor
XTL	198 kHz transmitting crystal
Loop	See text
M1	0-50 μA dc meter
R1	220 kilohm 1/4-watt 5% resistor
R2	1 kilohm 1/4-watt 5% resistor
R3	2.2 kilohm 1/4-watt 5% resistor
S1	SPST toggle switch
Miscellaneous	Cabinet, hardware, etc. (see text)

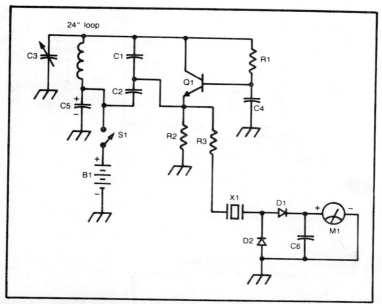

Fig. 2-23. Schematic diagram for single transistor locator.

TWO-TRANSISTOR-BFO LOCATOR

Building the two-transistor circuit requires that you follow the basic mechanical and electrical practices used in the construction of the first three locator projects. You should use the same basic layout as used in the other BFO circuits. Table 2-7 and Fig. 2-24 contain the parts list and schematic.

Table 2-7. Parts List for Double Transistor Locator.

B1	9-volt transistor battery
C1, C2	2720 pF/100-volt disc ceramic capacitor
C3, C6, C7	.01 μF/100-volt mylar capacitor
C4, C5	680 pF/100-volt disc ceramic capacitor
C8	47 μF/16-volt electrolytic capacitor
L1	Search loop (see text for details)
L2	Broadcast loopstick (tunable)
Headphones	2 kilohm headphones
Q1, Q2	2N3904 npn silicon transistor
R1, R3	220 kilohm 1/4-watt 5% resistor
R2, R4, R5	1 kilohm 1/4-watt 5% resistor
Miscellaneous	(See text)

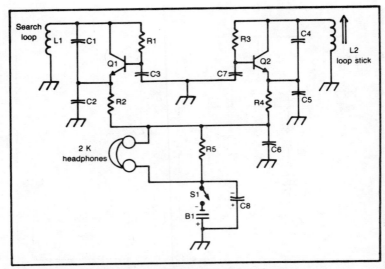

Fig. 2-24. Schematic diagram for double transistor locator.

To make the job of building the loop somewhat easier, follow the drawings in Fig. 2-25. The Faraday shield is made out of a section of 3/8-inch copper tubing formed into a 10-inch circle. Locate

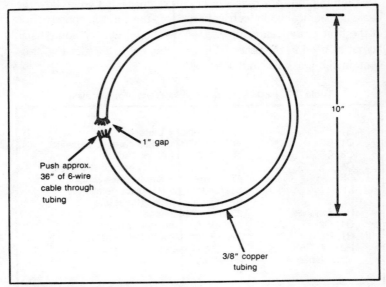

Fig. 2-25. Construction of 10-inch loop.

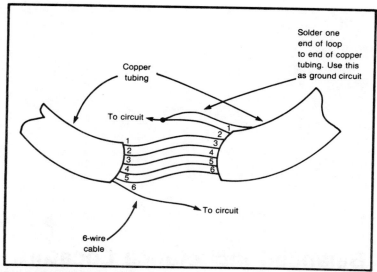

Fig. 2-26. Loop wiring.

about three feet of 6-wire antenna rotator or intercom cable, and fish it through the circle of copper tubing. Connect the wire ends together as in Fig. 2-26, and you will end up with a continuous 6-turn loop. One end of the loop is connected to one end of the copper tubing, and this common connection will go to the ground or positive battery circuit.

The operation of the circuit is as simple as the circuit itself. Q1 and it's associated components form a Colpitts oscillator with the loop as the tuned inductor. Q2 is connected in almost the same Colpitts oscillator circuit but with a loopstick coil as the tuned inductor. The emitters of Q1 and Q2 share a common path to the negative battery circuit. The mixed audio signal is taken off at the junction of R2, R4, and R5. A sensitive pair of high-impedance phones are used so the volume level will be sufficient. To keep the two oscillators from locking together, a .01 µF capacitor is used to bypass the rf at the resistor junction and allow the audio to pass to the headphones. Tuning the BFO locator is easy; just turn the loopstick's core until a beat note is heard, and proceed as with any other BFO locator.

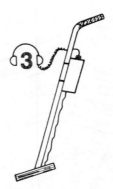

Balanced Inductance Locators

The most interesting of the balanced inductance locators is the balanced bridge detector. The search loop is placed in an inductance bridge circuit, and the bridge is electronically balanced so no output is present when the search loop is clear of any metal objects. When a metal object is placed in the field of the search loop this delicately balanced condition is shifted slightly and this small change upsets the total balance of the bridge circuit. A very small signal is then present at the output of the bridge and this signal is then amplified many times to drive an indicator circuit. The balanced bridge locator can be built to search out small or large metal objects, by selecting the size of the loop and by using the proper circuit design.

TWO-COIL BALANCED-BRIDGE LOCATOR

Figure 3-1 shows the first balanced bridge locator. The locator is constructed in two separate units. The dual loop and oscillator circuits make up the bottom half, and the high-gain amplifier and indicator circuits make up the upper half of the locator.

Refer to the circuit diagrams in Fig. 3-2 and Fig. 3-3. The complete parts lists are shown in Tables 3-1 and 3-2. The search loop coils, L1 and L2, and the associated components make up the Colpitts oscillator circuit. Both of the coils are identical in size and in the number of turns, and are mounted one on top of the other with the windings connected in an additive field configuration (the coils

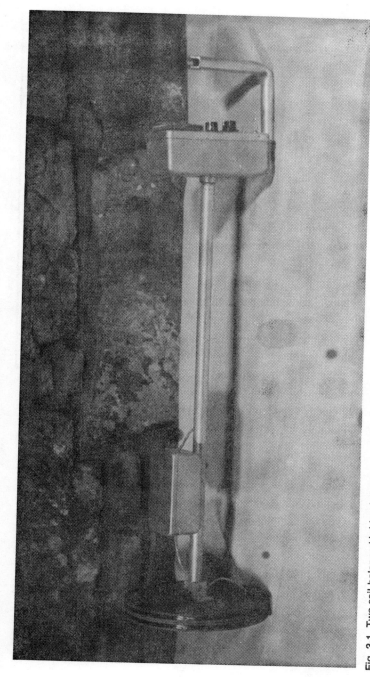

Fig. 3-1. Two-coil balanced-bridge locator.

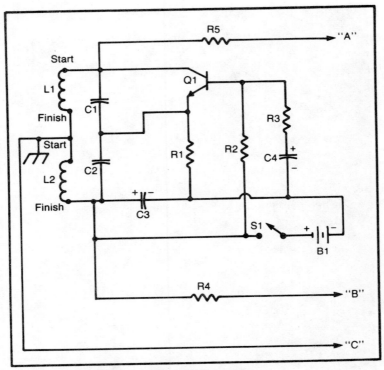

Fig. 3-2. Oscillator circuit of two-coil B/B locator.

spiral the same direction). The phase of the signal at the junction
of Q1 and L1 is 180 degrees out of phase with the signal at the
junction of C2 and C3. Both of the signals are the same level. The
junction of the two coils is connected to common, or circuit ground.
The three circuit connections between the oscillator circuit and the
amplifier carry the phase-related signals that are processed to indicate the presence of metal. A quad-op-amp IC supplies a total
circuit gain of 100,000 to amplify the unbalanced signal to drive
the meter and audio circuit.

A simplified diagram is given in Fig. 3-4. The signal at the top
of L1 and the bottom of L2 are equal in level and opposite in phase.
A high-gain amplifier is connected to the common coil connection
and to the arm of a balance pot, R_C. Without any metal present,
the circuit is balanced by setting the R_C control to a point where
the two out-of-phase signals are equal in value. At this point there
will be no signal present at the input of the amplifier. Unbalance
either of the coils and the difference in signal level will show up

40

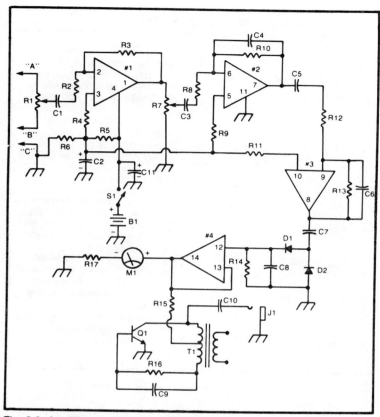

Fig. 3-3. Amplifier circuit of two-coil B/B locator.

Table 3-1. Parts List for Oscillator Circuit.

B1	9-volt transitor battery
C1, C2	.87 µF/100-volt mylar capacitors (.27 µF + .27 µF + .33 µF)
C3	330 µF/16-volt electrolytic capacitor
C4	4.7 µF/16-volt electrolytic capacitor
Q1	2N2924 npn silicon transistor, or 2N2222
L1, L2	Detector loops 1 and 2 (see text for details)
R1	1 kilohm 1/4-watt 5% resistor
R2	100 kilohm resistor
R3, R4, R5	10 kilohm resistor
S1	STSP toggle switch
Miscellaneous	Cabinet, wire, hardware, etc., (see text for details)

Table 3-2. Parts List for Amplifier Circuit.

B1	9-volt transistor battery
C1, C3, C5, C7, C8, C9, C10	.1 μF/100-volt mylar capacitor
C2	47 μF/16-volt electrolytic capacitor
C4, C6	39 pF/100-volt disc ceramic capacitor
C11	330 μF/16-volt electrolytic capacitor
D1, D2	1N914 silicon diode
Q1	2N2924 npn transistor, or 2N2222
IC-1	LM324 quad op amp
J1	1/4-inch phone jack
M1	0-50 μA dc meter
R1	5 kilohm 10-turn linear taper pot
R2, R4, R14	10 kilohm 1/4-watt 5% resistor
R3, R16	100 kilohm 1/4-watt 5% resistor
R5, R6	4.7 kilohm 1/4-watt 5% resistor
R7	2 kilohm taper pot
R8, R9, R11, R12	2.2 kilohm 1/4-watt 5% resistor
R10, R13	220 kilohm 1/4-watt 5% resistor
R15	1 kilohm 1/4-watt 5% resistor
R17	91 kilohm 1/4-watt 5% resistor
S1	SPST toggle switch
T1	8 ohm to 1 kilohm (with center tap) miniature audio transformer
Miscellaneous	Cabinet, hardware, wire, etc.

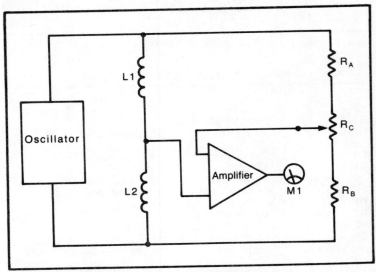

Fig. 3-4. Simplified diagram of two-coil B/B locator.

at the input of the amplifier. The amount of the unbalance will be indicated by the meter, M1.

An unusual feature of this locator circuit is the use of the search loop coils as the inductor for the oscillator circuit, similar to BFO locators. The design method most often used in this type of locator circuit has a separate oscillator, a buffer, and a driver amplifier to supply the search loop with its signal. The advantage of the single circuit is that it not only saves parts, but also requires much less battery power. The other major difference is the use of two coils in the bridge circuit, as opposed to the more complicated Maxwell inductance bridge that can be very difficult to balance to a zero signal level.

Building the Two-Coil Locator

Start construction by making the two search loop coils as in Fig. 3-5. Cut two 9-inch circle forms from pressed wood or any other easy to work, nonmetallic material 1/2 inch in thickness. Cut a 1/4- x -1/4-inch groove in the outer edge of each coil form. Use a

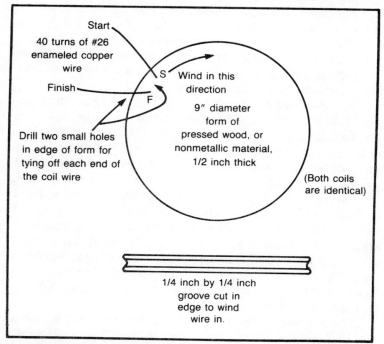

Fig. 3-5. Construction of the two-coil loop.

router or any other method that's convenient for cutting the groove. Drill two small holes, about 1 inch apart near the edge of the form and through one side of the 1/4-inch lip. The holes will be used to tie off the ends of the coil wires.

Wind 40 turns of number 26 enameled copper wire, in the direction shown in the diagram, on each of the forms. The actual direction of winding is not important, as long as both coils are wound in the same direction. Position one coil on top of the other so that the windings are going in the same direction. Glue them together and clamp in place. Connect the coil wires together as shown in the schematic of the oscillator. Hook the "finish" wire of the top coil to the "start" wire of the bottom coil. Mount a small three-terminal barrier strip near the edge of the top coil close to where the coil wires are located. Connect the "start" wire of the top coil to an outside terminal and the "finish" wire of the bottom coil to the other outside terminal. Hook the common coil wires to the center terminal and set the search loop assembly aside for now.

The Colpitts oscillator circuit is housed in a small 2- × -3- × -5 inch aluminum cabinet. Any other similar metal cabinet will do just as well. The exact size and shape of the cabinet is not important, as long as all circuit components will fit in place without crowding. The drawing in Fig. 3-6 shows the major parts layout. A section of 2 3/4- × -3 1/2 inch perf-board is used to mount the circuit components. Just follow the general layout shown, or use any convenient scheme that fits your needs.

The completed oscillator cabinet is mounted on a 26-inch section of aluminum conduit, 4 inches up from the bottom end. Two small aluminum "L" brackets are used to attach the search loop to the conduit. A 1 1/2-inch 6-32 screw, two flat washers, and a wing-nut holds the search loop in place. By loosening the wing-nut the search loop can be set to any desired angle and locked in place.

The amplifier and indicator circuits are housed in a 7- × -3- × -2 1/2-inch aluminum cabinet. The circuit components are mounted on a section of 2 3/4- × -5 1/4-inch perf-board. For a general parts layout follow the one shown in Fig. 3-7. The actual layout is not critical and any good scheme will work just fine. Keep all wire leads short, and mount all components with flea clips. The meter, R1, R7, S1, and J1 are mounted on the front panel, as shown, but any arrangement should do as well.

The perf-board is mounted to the back of the front panel with four 1 1/2-inch aluminum spacers (see Fig. 3-8). The metal cabinet is mounted to the top of the aluminum conduit, and a handle shaped

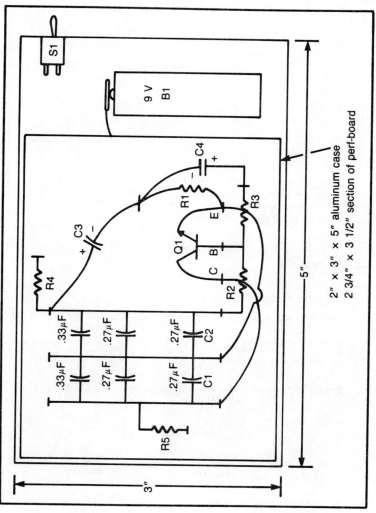

Fig. 3-6. Parts layout for the oscillator circuit.

2" × 3" × 5" aluminum case
2 3/4" × 3 1/2" section of perf-board

45

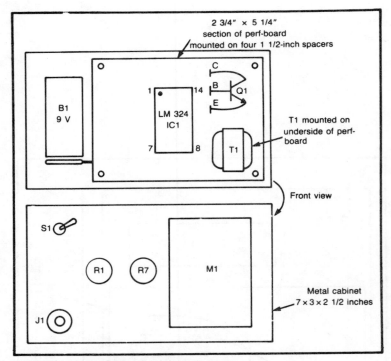

Fig. 3-7. Parts layout for the amplifier circuit.

from a short section of the conduit is mounted to the back of the cabinet as in Fig. 3-1.

Testing and Operating the Locator

Place a 9-volt battery in both circuits, and turn on the power

Fig. 3-8. Inside of amplifier circuit.

46

to the oscillator circuit. Set the ten-turn pot, R1, to mid-range and set the gain control R7 about 1/4 up from the minimum setting. If you turn S1 on the meter should read near full scale, and a tone will be present in the headphones. Clear the search loop of any metal objects, and turn R1 either direction until the meter reading drops to a null (minimum) reading. At this setting turning R1 in either direction should cause the needle to rise sharply. The most sensitive setting will occur when the gain pot R7 is set to maximum and a meter null is obtained with the ten-turn pot, R1. To achieve the best possible null or balanced condition a small piece of printed circuit board material (single sided), 1/2 × 1 1/2 inches in size can be used to balance out the metal used in the search loop construction. Set the locator for the best possible balance with the above procedure using R1 and R7.

Remove all watches, rings and any other metal items that could interfere with the final balancing of the search loop. Position the search loop away from any metal and take the small section of printed circuit board with the copper side toward the bottom side of the search loop, and slowly move it around near the edge of the loop until the deepest null is found. It may be necessary to readjust the balance pot while finding the best spot for the circuit board material. When the exact spot is found glue the board in place. Be prepared to spend some time obtaining the best possible null, as the overall performance will depend on how good the balance is.

If the locator doesn't respond properly, or refuses to operate at all, check the voltage readings given in Table 3-3. All voltage readings are taken with a 20 kilohm-per-volt meter with the nega-

Table 3-3. Voltage Readings for Two-Coil B/B Locator.

Circuit Location	Voltage Reading	Notes
Collector Q1 (TX)	9 volts	Common meter point at negative battery
Base Q1 (TX)	6 volts	Common meter point at negative battery
Emitter Q1 (TX)	6 volts	Common meter point at negative battery
IC1 LM324		With oscillator circuit off
Pin #1	4.5 volts	With oscillator circuit off
Pin #7	4.5 volts	With oscillator circuit off
Pin #8	4.5 volts	With oscillator circuit off
Pin #14	1.5 volts	With oscillator circuit off
Collector Q1 (REC)	1.9 volts	With oscillator circuit off

tive lead connected to circuit ground. Voltage readings within ± 20% are normal. The two-coil locator will offer the advanced experimenter the opportunity to work with different search coils with regard to size and shape, and to make minor circuit changes.

The gain of each op amp can be varied by changing the gain setting resistor values. The gain of each op amp is set by the ratio of the input resistor and the feedback resistor. In the first amplifier stage the input resistor (R2) is 10 kilohms and the feedback resistor (R3) is 100 kilohms. The value of R3 divided by R2 gives a gain of ten for the first amplifier stage.

The second gain stage has an input resistor (R8) of 2.2 k and a feedback resistor (R10) of 220 k, giving a gain of 100 for this op amp. The third amplifier stage uses the same values of resistors and also has a gain of 100. To calculate the total gain of the three amplifier stages, the gain of the first stage is multiplied by the gain of the second, and that product is multiplied by the gain of the third. The resulting total is the gain of the full three-stage amplifier. Thus, the total gain is G1 × G2 × G3 = 10 × 100 × 100 = 100,000.

If the feedback resistor of the first stage (R3) is doubled in value the gain is increased to 20, and the total gain of the three stages is increased to 200,000. If the feedback resistor (R3) is cut in half the gain is reduced to 5 and the total gain is reduced to 50,000. If an excellent null condition is obtained, the gain of the first stage can be increased to take advantage of the maximum sensitivity of the balanced-bridge circuit.

The search loop can be changed in size and shape to meet special requirements or just for the fun of experimenting with different designs.

If you choose to change the size of the loop coils, try to keep about 95 feet of wire on each coil. No matter what size coils are used this amount of wire will keep the coil's inductance within the range needed for it to function properly in this circuit. The two coils can be separated from each other, either vertically or horizontally to cover twice the usual search area.

The two-coil design offers a number of circuit variations to experiment with, but it would be useful to first obtain suitable operation with the circuit as shown and then expand from that point.

FOUR-COIL BALANCED-BRIDGE LOCATOR

The four-coil balanced-bridge metal locator is designed with the serious electronic hobbyist in mind (see Figs. 3-9 and 3-10). One

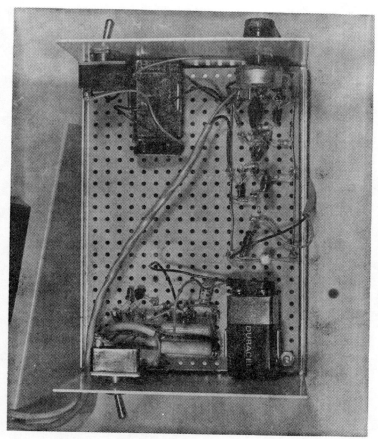

Fig. 3-9. Inside of the four-coil B/B locator.

very important feature of this locator is that the search area is four times that of a single-coil locator, and each of the four coils is as sensitive to small objects as any individual single-loop locator. The four-coil locator operates as if four single-coil locators were tied together and being used as a single unit. Of course, four separate locators would interfere with each other and would not work at all.

Circuit Description and Operation

The four-coil balanced-bridge circuit is considerably different from any locator we've covered up to this point, so a complete circuit and operation description will be helpful in fully understand-

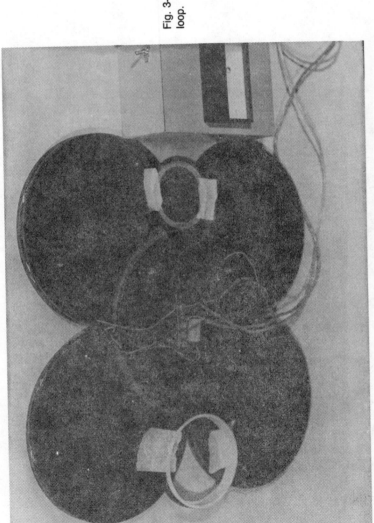

Fig. 3-10. Four-coil B/B search loop.

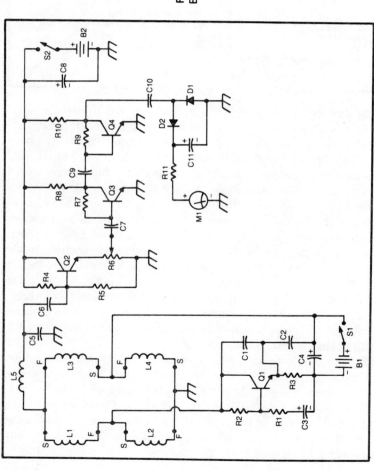

Fig. 3-11. Circuit diagram of the four coil B/B locator.

51

Table 3-4. Parts List for Four-Coil B/B Locator.

B1	9-volt transistor battery
C1, C2	.54 μF/100-volt mylar capacitor
C3, C11	4.7 μF/16-volt electrolytic capacitor
C4, C8	47 μF/16-volt electrolytic capacitor
C5	.27 μF/100-volt mylar capacitor
C6, C7, C9, C10	.1 μF/100-volt mylar capacitor
D1, D2	1N914 silicon diodes
Q1, Q2, Q3, Q4	2N2924 npn transistor, or 2N2222
L1-L5	Search loops and balancing loop (see text)
M1	0-200 μA dc meter
R1	10 kilohm 1/4-watt 5% resistor
R2	100 kilohm 1/4-watt 5% resistor
R3, R11	1 kilohm 1/4-watt 5% resistor
R4, R5	220 kilohm 1/4-watt 5% resistor
R6	10 kilohm linear taper pot
R7, R9	470 kilohm 1/4-watt 5% resistor
R8, R10	4.7 kilohm 1/4-watt 5% resistor
S1, S2	SPST toggle switch
Miscellaneous	Cabinet, hardware, etc.

ing this unusual locator (see Fig. 3-11 and Table 3-4).

The search loop is made up of four identical coils and one small balancing coil. The phasing of the four 9-inch coils and the position of the balancing coil is shown in Fig. 3-12. Each of the four coils are of 40 turns each and are constructed like the coils used in the previous balanced-bridge locator described in Fig. 3-5. A fifth coil form is used to tie the four coils together mechanically as in Fig. 3-10. The coils are connected as shown with each start winding connected to a finish winding of an adjacent coil, making a four-terminal inductance bridge circuit. The center connections of the bridge circuit are connected in a Colpitts oscillator circuit, and C1 and C2 set the oscillation frequency to about 10 kHz. The opposite sides of the bridge circuit are connected to the input of a three-stage amplifier circuit. Transistor Q2 is connected in an emitter-follower circuit that isolates the output of the bridge from the loading of the two-stage transistor amplifier.

The output signal produced when the bridge is unbalanced by a metal object is amplified by two transistors, Q3 and Q4, to a voltage level sufficient to drive the indicator meter, M1. The sensitivity pot R6 sets the gain of the amplifier and is also helpful in making the initial balance of the four-coil search loop. A close look at the circuit diagram will reveal that there is no means of electrically balancing the search loop. The four-coil circuit is designed to be mechanically balanced only, and this is accomplished by the use of the fifth coil and a section of nonferrous metal. With a little patience and a few minutes the bridge can be balanced to an excellent null.

Building the Four-Coil Locator

Start construction by winding four identical 40-turn coils as shown in Fig. 3-5. Take extra care in keeping the "start" and "finish" wires plainly marked, and make sure all coils are wound in the same direction. Cut out another coil form, but don't bother cutting a groove in the outside edge as in the other four coil forms. This fifth coil form will be used to hold the four coils in place as in Fig. 3-10. Position the four-loop coils as shown in Figs. 3-10 and 3-12, so the four outside edges of the four coils are exactly square before mounting in place with the fifth coil form.

Four 8-32 screws, washers and nuts are used to hold the four

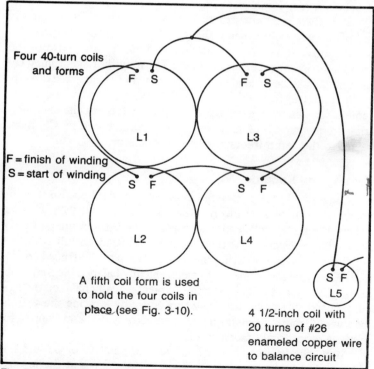

Fig. 3-12. Diagram of four-coil wiring hookup.

coils in place on the fifth form. Connect the coil wires at the start and the finish ends to match the wiring diagram shown in Fig. 3-12.

The coil form for L5 can be made of either wood or plastic, and should be 4 1/2 inches in diameter. Wind 20 turns of #26 enameled wire on the form in the same direction as the four search loop coils. Mark the start and finish ends of the coil and connect to L2 and L3 as shown in Fig. 3-12. Locate the 4 1/2-inch coil in the general location shown in Fig. 3-10. Tape the coil in place and set the loop assembly aside for now.

The electronic circuit for the locator is simple and straightforward and can be housed in any suitable metal case large enough to hold all of the components without crowding. Don't use a cabinet that is too small to work with, especially if you decide to experiment with the circuit later. The cabinet used is a two-piece aluminum case 7 × 5 × 3 inches in size. The oscillator circuit is located on one end of a 6 3/4- × -4 1/2-inch section of perf-board with the amplifier circuit on the opposite end with enough room in between for circuit expansion.

Start the circuit construction by building the oscillator at one end of the perf-board, using flea-clips to mount the components. A four-terminal barrier strip is mounted on the fifth form of the search loop to connect the coil wires. The amplifier loop tuning capacitor, C5, can be located on the barrier strip or connected directly to the circuit in the cabinet. Locate C5 where it is the most convenient for tuning or experimenting with different coils and operating frequencies. The amplifier loop section, tuned by C5, must be set to the same frequency as the oscillator.

Testing the Locator

Connect a battery to each circuit and set gain pot R6 to midposition. Turn on the power to the amplifier circuit (S2). If all is normal the meter will read near zero. If you turn on the power to the oscillator circuit (S1) the meter should jump to full scale.

The next step is to bring the four-coil loop to a balanced condition. Since there is no means to electronically balance the four-loop bridge circuit, a special procedure to mechanically balance the circuit is used. Start the balancing procedure by locating the search loop away from any metal object and setting the gain control to produce a half-scale meter reading.

Move the balancing coil, L5, over the junction of L3 and L4 (refer to Fig 3-10). The best balance adjustment is at the lowest meter reading. As perfect balance is approached, the amplifier's gain (R6) must be increased to keep the meter reading near one-half scale so the null can be easily obtained. When the best location for L5 is found, temporarily tape the coil in place and procede to the final balancing step.

The final balancing step is to put a piece of ferrous metal in a location on the search loop that will bring the bridge circuit into a nearly perfect state of balance. A suitable metal object that can be found almost anywhere is a wide-mouth Mason canning jar lid. The diameter of the jar lid is 3 1/4 inches. If a jar lid can not be found, try using a similar size piece of nonferrous metal as a substitute. Start with the jar lid in an area near the junction of L1 and L2 (see Fig. 3-10) and slowly move the lid around in that general area until you find the best null. Once you find the best location of the lid, an even better balance might be obtained by repositioning L5. The best condition is obtained by "juggling" the locations of L5 and the jar lid. When you have found the optimum null, glue coil L5 in place. The jar lid can also be kept in place with glue, but if you want to experiment with the locator, a good quality tape would be the best choice for holding the lid in place.

Using the Locator

The four-coil search loop is too heavy to carry for any period of time, so there is no handle in the photo of the locator. A handle with a counter-balance weight might be a good way to go, but no matter what method is used to carry the search loop, the apparatus should be constructed from wood or plastic, since any metal items will cause an unbalanced condition in the search loop.

One unusual application of the four-coil search loop locator is to mount the search loop out in front of a motor vehicle. With this set-up a large area can be covered in a short period of time, but the probability of finding any small objects is unlikely at best. The potential and versatility of the basic four-coil locator is limited only by your own imagination. You should feel free to experiment with your own ideas until you get the results you want.

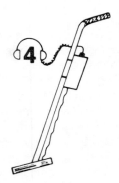

Transmitter/Receiver Locators

The Transmitter/Receiver method of metal detection is different from that used by the frequency-shift or the balanced-bridge locator circuits. An rf field is emanated from the transmitter loop and the receiver loop is positioned in a null or a near-zero signal area of the field pattern. Under the balanced-field condition the receiver's loop does not see any of the transmitter signal and no metal is indicated. When a metal object is moved into the transmitter's field, a new pattern is created around the metal object that upsets the null condition of the receiver loop. The receiver circuit amplifies the rf signal and indicates the presence of metal by a meter reading or an audio tone.

The transmitter/receiver (T/R) metal locator can be designed to locate either as small coin-sized object or to seek out a larger metal object buried several feet deep. A single T/R metal locator can not perform both tasks. The T/R locator designed to locate large metal items must use large loops for the transmitter and receiver, and the large loops will not see any metal object smaller than a silver dollar. The T/R that is designed to locate small objects requires the use of small loops for the transmitter and receiver, and the small size of the loop limits the depth of detection to just over one foot for almost any size object.

The T/R metal locator shown in Fig. 4-1 is designed to locate small coin-sized objects, and will detect a dime four to five inches from the bottom of the search loop in open air. The actual depth

Fig. 4-1. Transmitter/receiver coin-shooter locator.

of detection of a similar metal object underground will depend on the soil condition and the length of time the object has been buried. Normally, the longer a metal object remains underground the easier it is to locate. The chemical action of water and oxygen on most buried metal objects causes the ground to absorb the minute metallic particles that migrate from the object, giving a larger target for the locator to see. Buried iron items tend to give larger target area much sooner than most other metals due to the fast oxidizing or rusting of iron.

T/R COIN-SHOOTER LOCATOR

The schematic diagram in Fig. 4-2 and parts list in Table 4-1 give an idea what component parts you need to build the T/R coin-seeker metal locator. The transmitter circuit is composed of the loop coil, L1, Q1, and associated components. The receiver section consists of tuned loop L2, buffer amplifier Q2, and a two-stage amplifier, Q3 and Q4. The amplified signal is rectified to dc by D1 and D2. The dc output supplies a bias current that activates the audio oscillator, Q5, and associated components. The audio tone is fed to the headphones to indicate the presence of metal. After the construction procedures have been covered I will give a more complete description of the operation of the circuit.

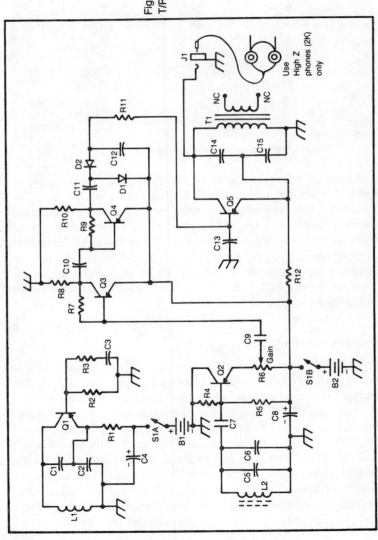

Fig. 4-2. Circuit diagram of T/R coin-shooter locator.

Table 4-1. Parts List for Coin-Shooter T/R Locator.

B1, B2	9-volt transistor battery
C1, C2, C14, C15	.27 μF/100-volt mylar capacitor
C3, C11, C12, C13	.1 μF/100-volt mylar capacitor
C4, C8	330 μF/16-volt electrolytic capacitor
C5, C7	.056 μF/100-volt mylar capacitor
C6	.015 μF/100-volt mylar capacitor
C9, C10	.047 μF/100-volt mylar capacitor
D1, D2	1N914 silicon diode
Q1-Q5	2N4249 pnp transistors
J1	1/4-inch phone jack
T1	1 kilohm to 8-ohm min. audio transformer
R1, R3, R12	1 kilohm 1/4-watt 5% resistor
R2, R4, R5, R9	220 kilohm 1/4-watt 5% resistor
R6	2 kilohm linear taper pot
R7	470 kilohm 1/4-watt 5% resistor
R8	4.7 kilohm 1/4-watt 5% resistor
R10	2.2 kilohm 1/4-watt 5% resistor
R11	100 kilohm 1/4-watt 5% resistor
L1, L2	Transmitter and receiver loops (see text)
Headphones	2 kilohm phones
S1	DPDT toggle switch
Miscellaneous	Cabinet, hardware, etc.

Building the Search Loop

The basic components for the search loop consist of a 4 1/2-inch diameter section of plastic pipe for the transmitter loop form, wood for end pieces, a 1/4-inch diameter section of a ferrite rod, and number 26 enamel-covered copper wire. Figures 4-3 and 4-4 show the following construction steps. The best place to start building the transmitter loop is to cut a 1 1/4-inch length of 4 1/2-inch plastic pipe. Wind a 60-turn coil on the plastic form with number 26 enamel-covered copper wire (see Fig. 4-4). Drill two small holes at each end of the coil winding and push the wire ends through to the inside of the coil form, and tape them in place.

Locate a piece of 1/4 inch diameter ferrite rod and cut to a 2 1/4-inch length. If finding a source for the ferrite rod is a problem, try using an old loop stick or antenna core salvaged from a junked transistor radio. Jumble-wind 150 turns of number 26 enameled wire, keeping within a 1 1/4-inch section of the middle of the ferrite rod as in Fig. 4-5. Tape the wire ends so the coil will not unwind.

Cut two wooden circles out of 1/4-inch material to fit snugly inside each end of the transmitter's coil form. Two hardwood blocks are shaped to hold the receiver's loop (L2) in place on the bottom circle of wood that serves as a mounting base (see Fig. 4-6).

The coil adjustment block should match the drawing in Fig.

Fig. 4-3. Inside view of coin-shooter's transmitter loop.

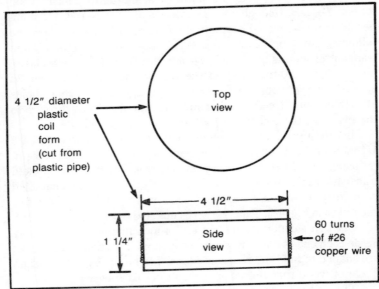

Fig. 4-4. Diagram of transmitter loop.

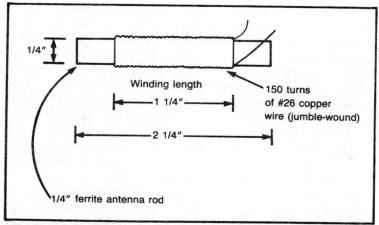

Fig. 4-5. Diagram of receiver loop.

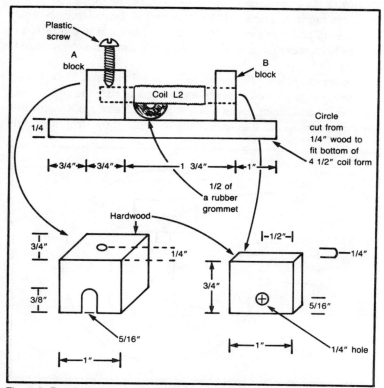

Fig. 4-6. Receiver coil placement and parts adjustment.

61

4-6, and should have a threaded hole to match the plastic screw used for the balancing adjustment. The actual size of the screw can be almost any size that is available, but a #8 to a 1/4-inch would be a good choice. Be extra careful in drilling and taping the hole in the wood, as it will split if extreme care is not taken. The second block serves to hold the fixed end of the receiver's coil and should match the drawing in Fig. 4-6. No metal is used in the search loop construction, because even a small 6-32 nut can prevent the search loop from balancing properly.

Take one of the circle cut-outs and glue it in the bottom of the 4 1/2-inch transmitter coil form. Allow ample time for the glue to set up before proceeding with the loop construction.

Position the receiver coil mounting block, "B", on the inside bottom of the search loop assembly and glue the block in place (see Fig. 4-7). Take the receiver loop and push one end of the ferrite rod through the hole in "B" block. The end of the rod should be flush with the end of the wood block. The fitting of the rod in the block can not be too tight, as the other end of the coil must be able to move up and down at least 1/8 inch. If the rod is too tight in the block, increase the hole size until the coil can move the required amount.

To mount block "A", refer to Figs. 4-6 and 4-7. Slide block "A" down over the coil's ferrite rod and make sure the coil is free to move up and down in the block's groove. Take a 3/8-inch rubber grommet and cut it in half. Take one half of the grommet and

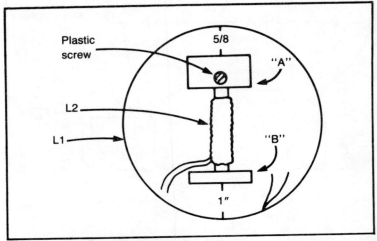

Fig. 4-7. Top view of search loop.

place it under the rod next to the "A" block as in Fig. 4-6. The rubber grommet will function as a return spring for the plastic balancing adjustment screw. When all parts are positioned in a workable location glue "A" block in place. After the glue has set, place the plastic screw in the block and check to see if the receiver coil will move up and down smoothly.

Cut the receiver and transmitter coil leads about 2 1/2 inches long. Remove the enamel cover from about 1/4 inch of each wire and tin the ends. Connect a 45-inch length of shielded microphone cable to the transmitter coil, and a similar length to the receiver coil. Solder and tape all wires. Do not connect the shields together at the search loop. This will cause problems in balancing the loop and reduce the overall sensitivity of the locator. Position the two shielded wires near one side of the transmitter loop and epoxy or glue them in place.

Take the remaining wood circle cut-out and drill three holes, one to allow the plastic balancing screw to pass through, one for the two shielded cables, and one in the center for a 1/2-inch wood dowel to serve as a handle for the search loop. Position the cut-out in the top of the transmitter coil form and glue to hold it in place. Glue or epoxy a full-length 1/2-inch wood dowel rod in the center hole to serve as a handle for the search loop.

Coin-Shooter Circuit Construction

The electronics for the T/R coin-shooter are housed in a 4- × -5 1/2- × -2-inch metal cabinet, but any metal enclosure of sufficient size can be used. For parts placement and general layout see Fig. 4-8 and 4-9. A section of perf-board 3 3/4 × 4 1/4 inches holds most of the components and is mounted to the bottom of the metal cabinet.

Start the circuit construction by locating and drilling all holes needed in the metal cabinet. Holes are needed to mount S1, R6, J1, perf-board, and the two phono jacks if used for the search loop connections. Flea-clips or push-in terminals are used to mount the electronic components and will also help keep everything stable mechanically. I recommend the point-to-point wiring method, but any proven scheme will do.

An excellent starting place in the circuit construction would be to locate the components for the oscillator circuit (TX) in the back left corner of the perf-board. The oscillator circuit can be completed and even tested before starting the receiver circuit construc-

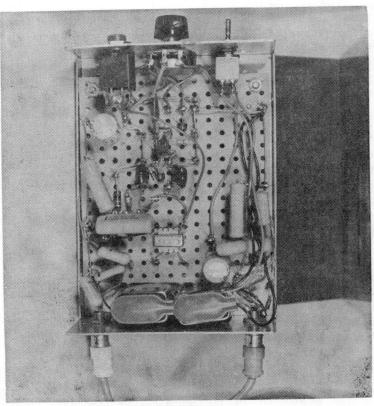

Fig. 4-8. Interior circuitry of the coin-shooter.

tion. If a frequency counter is available the oscillator frequency can be checked, and should be somewhere close to 20 kHz. The actual frequency is not critical if both the transmitter and receiver are tuned to the same operating frequency. If your construction of the search loop is close to the one shown there should be no need to check the actual frequency of oscillation.

For the receiver you can follow the general layout shown in Fig. 4-9, and start by placing the four transistors in their designated locations. After you complete the receiver wiring, check all connections and components to be sure that they follow the schematic diagram.

Checking Out the Circuit

Connect the search loop to the locator's circuit, and put the two

9-volt batteries, B1 and B2, in place. Plug a pair of high-Z headphones into J1 and turn S1 on. You should hear an audio tone in the headphones. Position the search loop away from any metal and set the gain control, R6, to about one-half full rotation.

Use a non metallic screwdriver and slowly adjust the plastic screw (balancing adjustment) up or down until you find a null in the audio tone. If the null or balance region is broad—one to two full turns—increase the receiver gain (R6) until the null is very sharp. If you can't get a very sharp null, then the receiver is probably not tuned to the transmitter's frequency. If electronic test equipment is not available, adjustments can still be successfully made in retuning the receiver's loop circuit to the transmitter frequency.

Set the balance adjustment screw for the best possible balance and bridge a small capacitor—a .001 µF will do to start—across C6. If the audio tone increases, then the receiver is tuned to a higher frequency than that of the transmitter, and a higher capacitance is needed at C6. Try replacing C6 with a .02 µF capacitor, and increase by .005 µF steps. If, after adding the capacitor, the audio tone decreases in level, the receiver is tuned to a frequency lower than the transmitter and a smaller capacitor is needed to replace

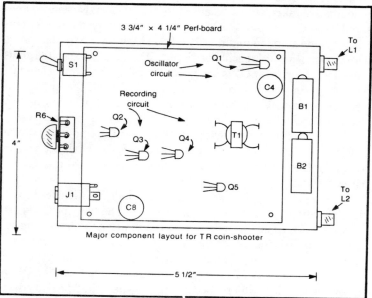

Fig. 4-9. Parts layout for the coin-shooter.

Table 4-2. Voltage Readings for Coin-Shooter T/R Locator.

Circuit Location	Voltage
Emitter of Q1	5.5 volts
Collector of Q3	3 volts
Collector of Q4	3 volts
Emitter of Q2	4 volts
Emiter of Q5	6 volts

C6. Start out by using a .01 μF capacitor to replace C6. When you have obtained the loudest tone level the receiver is tuned to the transmitter frequency and is ready to try out.

After the receiver has been tuned, repeat the balancing procedure and set the receiver's gain to the highest useable setting. Take a dime and bring it toward the middle area of the bottom of the search loop. The locator should emit a tone when the line is 4 to 5 inches from the search loop.

If the locator refuses to operate at all or functions only partially, refer to the voltages given in Table 4-2. All voltages are taken with a 20,000-ohms-per-volt meter. The common or negative meter lead connects to the circuit common, or battery negative. All readings are taken with a balanced loop condition. Measurements should agree to within $\pm$ 20%. If the trouble persists, the following circuit description might be helpful in locating the trouble. Start with the transmitter's oscillator circuit, Q1, and it's associated components. The bias for Q1 is set by R2, and should produce an emitter voltage of about 5 to 6 volts, measured from the emitter to chassis, or battery negative. Resistor R3 and capacitor C3 help shape the oscillator waveform to produce a near perfect sine wave.

Capacitors C1 and C2, along with L1, set the frequency of the Colpitts oscillator circuit. C4 guarantees a low-impedance power source for the oscillator, ensuring a stable operation even when the battery begins to deteriorate.

The receiver's tuned circuit, L2, C5, and C6 is isolated from circuit loading by the emitter follower, Q2. The bias for Q2 is set by resistors R4 and R5, which sets the emitter to about 4 volts. The gainpot R6, doubles as the emitter resistor and a variable signal level control. The rf signal is taken from the wiper of R6 and is fed to the base of Q3, the first gain stage. The bias for Q3 is set by resistor R7, which is connected to the collector and base, giving the amplifier stage a slight negative feedback. This type of bias adds to the overall stability of the amplifier. The second amplifier has the same type of base bias through resistor R9. The rf output is taken off at the collector of Q4 and is fed to a voltage-doubler rectifier circuit using diodes D1 and D2. The dc output of the voltage-doubler circuit is fed to the base of transistor Q5. Q5, T1, C14, and C15 make up a Colpitts audio oscillator circuit that feeds the phones through J1. As the dc output increases, the bias current fed to the base of Q5 is raised enough to start the audio oscillator. The audio level varies with the unbalancing of the transmitter's field as metal is moved into the locator's detection range.

Using the T/R Coin-Shooter

The T/R coin shooter is designed to ferret out buried or hidden coin-sized metal objects. The best place to hunt, of course, is outdoors. Choose a place to hunt where many people congregate and where money is continuously changing hands. Playgrounds, fairgrounds, outdoor flea markets, school grounds, parks, and old homesteads are all excellent locations for using the T/R coin shooter. Always obtain permission no matter where you hunt and be courteous to all landowners.

To set the T/R coin shooter to the most sensitive operating mode, adjust the balance screw until a low audio tone is heard in the headphones. This mode of operation can be found with two settings of the balance screw. One mode will occur going in or clockwise with the adjustment screw as the balance point is approached, and the other mode is found by going past the balance point. In one of the modes the locator will respond with a louder tone for a ferrous metal object and a lower tone for a nonferrous object. Knowing which operating mode you are in can help you distinguish valuable from nonvaluable items. You can usually ignore all ferrous signals from the locator when coin shooting, and save yourself considerable digging time.

T/R DEEP-SEARCHER METAL LOCATOR

The deep-searcher T/R metal locator is shown in Fig. 4-10. The first impression after looking over the photo might be that the deep searcher is too complicated or expensive to build. Not so. The two box metal locator is the simplest of the two basic T/R locators. The single-transistor oscillator that functions as the transmitter is housed in one box, and the four-transistor receiver is housed in another box.

Fig. 4-10. Deep-searcher T/R locator.

The circuit diagram is shown in Fig. 4-11. It is a single-transistor Colpitts oscillator with the loop coil as part of the tuned circuit. Table 4-3 shows the parts list. The function of the transmitter circuit and it's operation is identical to the oscillator circuit used in the T/R coin-shooter locator covered previously.

The receiver circuit has a tuned-loop input followed by a three-stage transistor amplifier. Figure 4-12 and Table 4-4 give the schematic and parts list for the receiver. The output signal from the amplifier feeds a voltage-doubler circuit, D1 and D2. The dc output is coupled to the base of an emitter-follower transistor iso-

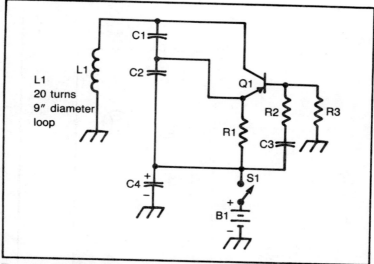

Fig. 4-11. Circuit diagram for the transmitter.

lation stage (Q4) that drives the 0-1 milliammeter. The meter indicates the signal level received from the transmitter when a metal object is within range.

Building the T/R Deep-Searcher Locator

Since both the transmitter and receiver circuits use an identical loop coil, start construction by winding two 9-inch 20-turn coils. The 9-inch loop forms are exactly the same as the 9-inch forms used in Chapter 2 and 3 and should be made the same way. Each loop has 20 turns of number 26 enamel-covered copper wire wound in the form's groove. The winding direction is not important with this type of T/R locator.

Table 4-3. Parts List for Deep-Searcher T/R Locator Transmitter.

B1	9-volt transistor battery
C1, C2, C3	.12 µF/100-volt mylar capacitor
C4	100 µF/16-volt electrolytic capacitor
Q1	2N4249 pnp transistor
S1	SPST toggle switch
L1	See text for materials
R1, R2	1 kilohm 1/4-watt 5% resistor
R3	220 kilohm 1/4-watt 5% resistor
Miscellaneous	Cabinet, hardware, etc.

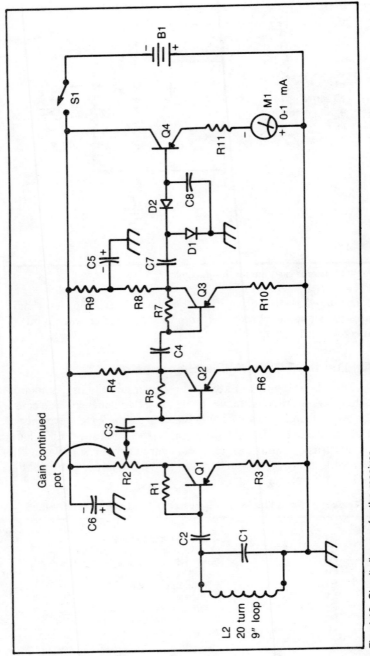

Fig. 4-12. Circuit diagram for the receiver.

70

Table 4-4. Parts List for Deep-Searcher T/R Locator Receiver.

B1	9-volt transistor battery
C1, C7	.056 μF/100-volt mylar capacitor
C2, C3, C4	.02 μF/100-volt mylar capacitor
C5	47 μF/16-volt electrolytic capacitor
C6	100 μF/16-volt electrolytic capacitor
C8	.1 μF/100-volt mylar capacitor
D1, D2	1N914 silicon diode
Q1-Q4	2N4249 transistors
M1	0-1 μA dc meter
R1	220 kilohm 1/4-watt 5% resistor
R2	2 kilohm linear taper pot
R3, R6	47-ohm 1/4-watt 5% resistor
R4	4.7 kilohm 1/4-watt 5% resistor
R5, R7	470 kilohm 1/4-watt 5% resistor
R8, R11	2.2 kilohm 1/4-watt 5% resistor
S1	SPST toggle switch
R9	1 kilohm 1/4-watt 5% resistor
R10	100-ohm 1/4-watt 5% resistor
Miscellaneous	Cabinet, hardware, etc.

The transmitter's plastic cabinet is mounted to the loop form with two 6-32 screws adjacent to the two coil leads (see Fig. 4-10 and 4-13). The transmitter circuit is built on a 3 1/2- × -1 3/4-inch section of perf-board, and is located in the transmitter cabinet as in Fig. 4-13. The circuit is not very critical and almost any good layout scheme can be used in its construction, but the layout in Fig. 4-13 might be the simplest to use. The completed transmitter and coil assembly mounts to the wood handle with a single wood screw as in Fig. 4-14.

The receiver circuit is housed in a plastic cabinet, and the components are mounted on a 5 × 2 3/4-inch section of perf-board. The drawing in Fig. 4-15 gives the basic parts layout for the receiver circuit. The meter and battery holder are mounted to the front of the plastic receiver cabinet, and the completed assembly is mounted to the loop form with two 6-32 screws adjacent to the coil wires.

The receiver and coil assembly is positioned on the opposite end of the wood handle in a manner that allows an up and down adjustment of the receiver to bring the T/R locator into balance. The hinged "U" bracket in Fig. 4-14 is made from a piece of scrap aluminum. The balance bracket can be made from the same material as the hinge and should match the drawing in Fig. 4-14. The U bracket mounts on the back of the coil form and the balance bracket on the front. A 2 1/2 inch 8-32 screw serves as the balancing adjustment, and a wing nut and washer and small spring sit

Fig. 4-13. Interior view of the transmitter.

on top of the wood handle with the balancing bracket. The coil spring should be stiff enough to keep the receiver sitting stable after the balance adjustment has been completed.

Putting the Deep Searcher Into Operation

Place a battery in each unit and switch on the power. Move away from any metal objects and turn the gain of the receiver to maximum. If all circuits are operating as they should, the meter should be at full scale. Reset the receiver gain (R2) until the meter reads a division or two below full scale, then turn the wing nut up and down until you have a null or minimum reading on the meter. When the receiver is properly balanced, the gain control should be able to operate at the maximum or near maximum setting. This is the most sensitive mode of operation for the T/R locator.

If the locator fails to perform as stated, then you may find the following troubleshooting information useful.

Ideally, the transmitter should operate at approximately 40 khz, but any frequency within ± 20% will work. No matter what the actual transmitter frequency is, the most important thing is to tune the receiver to the same operating frequency. If the receiver is not properly tuned, the sensitivity of the locator will be very poor, and the locator may not work at all.

To check the tuning of the receiver, you should use the follow-

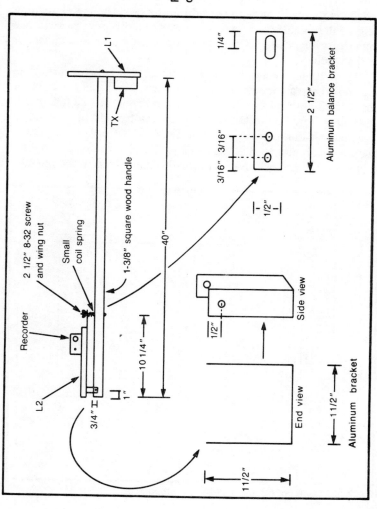

Fig. 4-14. Diagram of the deep-searcher locator.

ing procedure, which does not require a frequency counter or any other special test equipment. Move the locator away from any metal objects. Remove C1 from the receiver's tuned circuit and replace it with a .047 μF capacitor. Turn on both units and set the gain of the receiver to one-half scale. If necessary, unbalance the locator with the wing nut adjustment until the desired reading is obtained. Connect additional capacitance across the .047 μF capacitor, either with a capacitor decade box, or by tacking individual capacitors in place until the meter reading is at maximum. As the frequency of the receiver approaches the transmitter frequency the receivers gain will have to be reduced to keep the meter reading below full scale. If the receiver tunes with a different capacitance than the original C1 .056 mF, then replace it with a new capacitor of that value.

If the tuning is correct, but the locator still fails to work, then you should check the reference voltages listed in Table 4-5 for other possible circuit faults. All voltages are taken with a 20,000 ohms-per-volt meter. For voltages in the transmitter circuit, the negative meter lead is connected to the negative battery terminal, and the positive meter lead is used as a probe. For voltages in the receiver circuit, the positive meter lead is connected to the positive

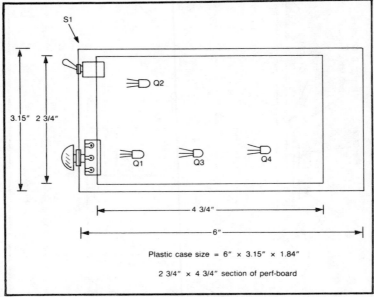

Fig. 4-15. Layout of the major receiver components.

Table 4-5. Voltage Readings for Deep-Searcher T/R Locator.

Circuit Location	Voltage
TRANSMITTER READINGS	
Emitter of Q1 in the TRANSMITTER circuit	+6 volts
RECEIVER READINGS	
Collector of Q1	−4 volts
Collector of Q2	−3 volts
Collector of Q3	−3.5 volts
Emitter of Q4	0 volts

battery terminal, and the negative lead is used as the probe. All receiver voltages are taken with the transmitter off.

Using the Deep-Searcher in the Field

The twin-box T/R locator is designed to seek out average to large metal objects that are buried as deep as three or more feet. A single silver dollar might be detected near the surface, but conditions have to be almost perfect to achieve this sort of sensitivity. Usually this type of T/R locator is used to find items no smaller than a one pound coffee can in size and shape.

Take the locator outside and adjust the balance of the receiver with the wing nut for a minimum meter reading. If a perfect balance is obtained, turn the receiver gain to maximum setting. If you can't obtain a perfect balance, set the gain to 1/4-scale meter reading. The deep searcher is ready for treasure hunting.

Hold the T/R locator at arm's length and parallel to the ground while walking over the search area. When the meter reading indicates a buried object, you can then approach it from several different directions to pin-point the buried object.

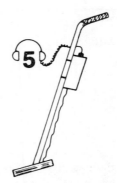

Coplanar VLF Locator

This chapter covers a single, but very sophisticated, coplanar induction-balance metal locator which is designed to locate both large and small buried objects. One unusual design feature is that the coplanar search loop replaces the basic two- and three-loop induction-balance search heads. This search loop design is relatively new and is used in a large number of commercial metal locators. It is ideal for both the T/R and induction-balance circuits.

The oscillator or transmitter feeds a single folded-loop transmitter coil that creates an induction-balanced field near the folded-back loop area. The receiver loop is parallel to the balanced-field area of the transmitter loop. Once the balanced position is located the receiver loop is mounted permanently with glue or epoxy. The transmitter field (with the folded-back loop) is self-cancelling for the receiver loop, but if a metal object comes into range this delicate balance is disturbed slightly and causes a small signal to enter the receiver coil. The detected metal can be indicated by a meter or by an audio signal, or both methods can be used.

The transmitter portion of the coplanar search loop is normally wound with several hundred turns of copper wire, with a small section of the loop folded back into a smaller loop to produce the self-cancelling field effect. This method of construction is a good one for the commercial builder, but it can be difficult for the hobbyist or home experimenter to duplicate on his own. In our locator the folded-back loop has been simplified, and requires only a single-turn loop in place of the multi-turn loop used normally.

The single-turn design is made possible by a special, easy to duplicate, hand-wound oscillator coil/transformer. By employing this special coil/transformer arrangement, the output impedance of the oscillator can be made to match the very low impedance of the single-turn loop. As in any electronic project, the matching of impedance is the single most important design criterion.

The operating frequency of the VLF locator is about 5.8 kHz, but the actual frequency can vary somewhat without affecting the operation of the circuit. Circuit construction methods at these low frequencies are not very critical, but the usual sound assembly practices should be followed.

CIRCUIT OPERATION

Starting with the oscillator circuit diagram in Fig. 5-1, and the parts list in Table 5-1, we can see that Q1 and its associated com-

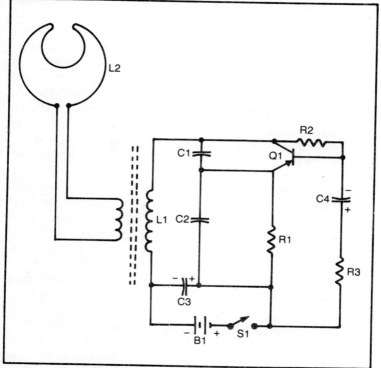

Fig. 5-1. Circuit diagram of the coplanar VFL oscillator.

77

Table 5-1. Parts List for Coplanar Locator Oscillator.

B1	9-volt transistor battery
C1	1 μF/100-volt mylar capacitor
C2	1.47 μF/100-volt mylar capacitor (1 μF and .47 μF connected in parallel)
C3	100 μF/16-volt electrolytic capacitor
C4	6.8 μF/16-volt electrolytic capacitor
L1	Oscillator coil with 4 1/2-inch- $\times$ -1/4-inch ferrite antenna rod (see text)
L2	Single-turn search loop (see text)
Q1	Silicon pnp transitor, 2N4249, 2N5086, or 2N3638A, etc. (Almost any general purpose small signal silicon pnp transistor will work)
R1	470 ohm 1/4-watt 5 or 10% carbon resistor
R2	100 kilohm 1/4-watt 5 or 10% carbon resistor
R3	3.3 kilohm 1/4-watt 5 or 10% carbon resistor
S1	SPST toggle or slide switch
Miscellaneous	See text for detail on items not listed here

ponents make up a standard Colpitts oscillator circuit. The oscillator frequency is set by the inductance of L1 and capacitors C1 and C2. L1 has a 200-turn primary winding as the oscillator's tuned inductance, and a four-turn secondary winding to match the single-turn loop. A 9-volt transistor-type battery supplies power for the oscillator circuit. The complete circuit is housed in a plastic cabinet.

The receiver circuit is shown in Fig. 5-2, and Table 5-2 shows the parts list. A 4 1/2-inch 50-turn coil serves as the tuned pick-up inductor, L3. C1 tunes inductor (L3) to the same frequency as the oscillator, 5.8 kHz. All active circuits used in the receiver are furnished by a single, quad op amp IC, LM324. The output of the tuned loop, is coupled to the input of op amp A, and the gain of $\times$ 10 is set by the values of R1 and R2. The gain for the three op amps is determined by a simple formula: gain = R2/R1. Output from amp A is connected to a 2 kilohm pot, R6, that functions as the locator gain or sensitivity control. Amp B boosts the 5.8 kHz signal 47 times (R8/R7), and the output is coupled to the input of amp C. The signal is multiplied another 21 times, giving a total overall gain of 9,870 times that of the small signal originally received by L3. Total gain is found by multiplying the gain of amp A $\times$ amp B $\times$ amp C, or 10 $\times$ 21 $\times$ 47 = total overall gain = 9870.

The 5.8 kHz output signal is fed to a voltage-doubler rectifier circuit using two silicon diodes, D1 and D2. Dc output from the doubler circuit is feed to op amp D, a voltage-follower stage, to prevent the meter from loading the rectifier circuit. The voltage gain

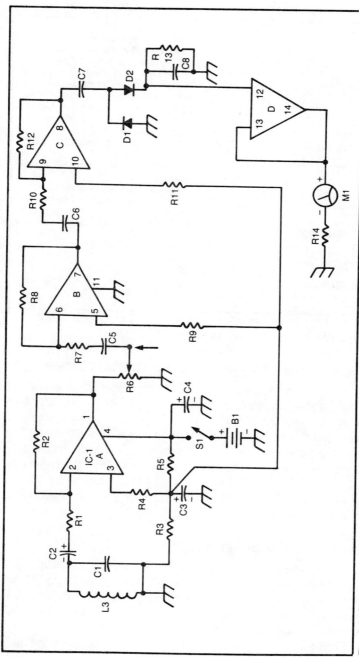

Fig. 5-2. Circuit diagram of the coplanar VFL receiver.

Table 5-2. Parts List for Coplanar Locator Receiver.

B1	9-volt transistor battery
C1	1 μF/100-volt mylar capacitor
C2	6.8 μF/16-volt electrolytic capacitor
C3, C4	100 μF/16-volt electrolytic capacitor
C5, C6, C7, C8	.1 μF/100-volt mylar capacitor
D1, D2	1N914 silicon signal diode
IC-1	LM324 quad op amp, 14-pin plastic IC
L3	See text
M1	0-200 μA dc meter
R1, R3, R4, R5, R7, R9, R10, R11	4.7 kilohm 1/4-watt 5 or 10% carbon resistor
R2	47 kilohm 1/4-watt 5 or 10% carbon resistor
R6	2 kilohm linear taper potentiometer
R8, R13	220 kilohm 1/4-watt 5 or 10% carbon resistor
R12	100 kilohm 1/4-watt 5 or 10% carbon resistor
R14	10 kilohm 1/4-watt 5 or 10% carbon resistor
S1	SPST toggle switch
Miscellaneous	See text for details on items not listed here

is, therefore, zero for this stage.

WINDING THE OSCILLATOR COIL/TRANSFORMER

The hand-made coil is wound around a section of ferrite antenna rod material 4 1/2 inches long by 1/4 inch in diameter (see Fig. 5-3 for winding details). Two rubber grommets are spread 2 inches apart in the middle of the rod material to set the width of the coil. Neatly wind 200 turns of number 26 enamel-covered copper wire between the two rubber grommets. Tape the coil wires to the ferrite rod at each end, and wind a four-turn coil around the middle of the 200-turn coil. The 200-turn coil goes to the oscillator circuit, and the four-turn coil connects to L2.

BUILDING THE FOLDED LOOP

Use a section of 1/8-inch copper tubing about 50 inches in length to create the loop as shown in Fig. 5-4. Locate the center of the 50-inch length of tubing and mark it with a small piece of tape. You can use L3's 4 1/2-inch plastic coil form as a winding guide for forming the 4 1/2-inch folded-loop section of L2.

Position the 1/8-inch tubing with the taped section in the middle of the 4 1/2-inch plastic form as in Fig. 5-4. Bend half of the 4 1/2-inch loop to the left of the tape mark, and the other half to the right of the tape mark. After forming the 4 1/2-inch circle, leave a 1/2-inch gap between the loop ends at the fold-back area. The 11-inch remaining loop can be formed around the 11 1/4-inch plywood search loop base and hand-shaped to just fit on top of the wooden base. If the loop ends overlap then trim off enough to allow for a 1-inch gap between the loop ends.

Drill 1/16-inch holes every 2 to 3 inches around the outside edge of the plywood base, and locate the holes 3/16 inch in from the outer edge of the base. Position the copper loop and mark around the 4 1/2-inch loop so holes can be drilled on each side of the loop. Lace the copper loop in place with small diameter nylon cord or fishing line. Glue or epoxy it for the best possible mechanical stability.

WINDING THE RECEIVER LOOP

A plastic coil form 4 1/2 × 1 inch is used to wind the receiver

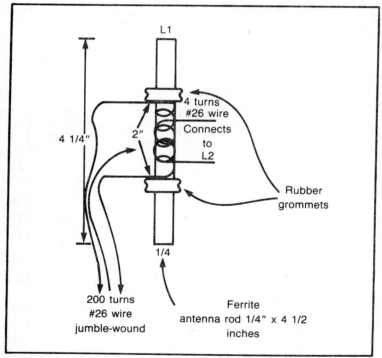

Fig. 5-3. Diagram of the oscillator coil/transformer.

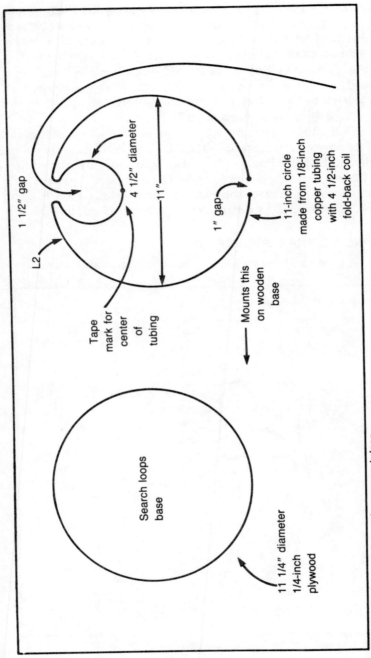

1 1/2" gap

4 1/2" diameter

11"

L2

1" gap

11-inch circle
made from 1/8-inch
copper tubing
with 4 1/2-inch
fold-back coil

Tape
mark for
center
of
tubing

Mounts this
on wooden
base

Search loops
base

11 1/4" diameter
1/4-inch
plywood

Fig. 5-4. Diagram of the coplanar search loop.

82

loop, and can be cut from a section of plastic pipe. Jumble-wind 50 turns of number 26 enamel-covered copper wire around the form and tape it in place (see Fig. 5-5).

BUILDING THE OSCILLATOR AND RECEIVER CIRCUITS

A photo of the complete locator is shown in Fig. 5-6, and a close-up of the coplanar search loop is shown in Fig. 5-7. The oscillator circuit is housed in a 2 7/8- × -4- × -1 1/2-inch plastic cabinet, and the basic component layout is shown in Fig. 5-8. A 3- × -1-inch section of perf-board holds most of the small parts, with the power switch (S1) mounted at one end of the cabinet and the ferrite rod coil at the other end. Two plastic cable clamps are used to hold the ferrite rod coil (L1) in place, and each clamp fits around the rubber grommet at each end of the coil. The two clamps are bolted to the plastic cabinet with 6-32 hardware. The 9-volt battery is

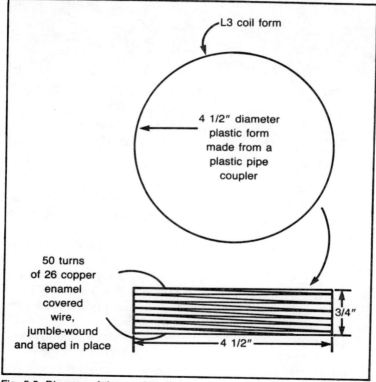

Fig. 5-5. Diagram of the receiver loop.

Fig. 5-6. Completed coplanar locator.

fastened to the inside of the cabinet with a small metal clip.

Almost any parts layout can be followed in building the oscillator circuit, since the operating frequency is very low and the total number of components is small. Keep all component wiring point to point and always double-check the wiring against the schematic diagram.

The receiver is housed in a 4- × -5 1/2- × -2 1/4-inch metal cabinet, and the basic part's layout is shown in Fig. 5-9. A 3 3/4- × -4 1/2-inch section of perf-board serves as the mounting board for the IC and most of the other small components.

The power switch (S1) and the receiver gain control (R6) mount on one end of the metal cabinet and the battery clip on the other. Indicator meter (M1) mounts to the top or lid of the metal cabinet, and the leads from the circuit to the meter should be made long enough for easy removal and servicing. As in the oscillator circuit, the parts layout isn't critical as long as a good wiring scheme is followed, and be certain that the wiring matches the schematic diagram.

A 36-inch piece of 3/4-inch diameter plastic pipe serves as a handle and mounting pole for all the components. (See the side view of the complete locator in Fig. 5-10). Mount the cabinet to the plastic pipe with 6-32 hardware 10 1/2 inches from the end of the handle. Drill a hole in the plastic pipe large enough for a twisted pair of number 22 wire, one inch behind the receiver cabinet. Run a length of the number 22 twisted-pair wire through the pipe and connect one end to the receiver input and the other end to the receiver loop coil (L3).

Mount the oscillator cabinet 18-inches from the handle's end with 6-32 hardware. Run a twisted pair of number 22 wire from the secondary of L1 to the two ends of copper tubing coil L2 and solder in place. The twisted pair connecting the oscillator to the loop should be wrapped loosely around the outside of the plastic pipe.

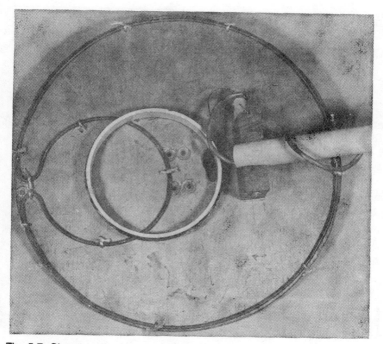

Fig. 5-7. Close-up of coplanar search loop.

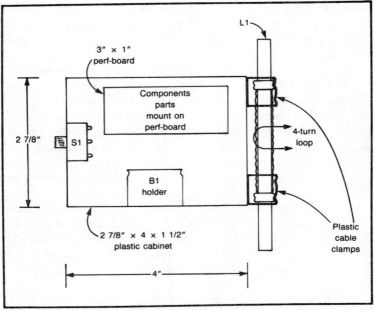

Fig. 5-8. Parts layout of oscillator.

COMPLETING AND TUNING THE COPLANAR LOCATOR

Make a simple wood hinge and mount it to the loop's plywood base as in Figs. 5-7 and 5-10, and drill a hole completely through the end of the pipe one-half inch from the end. A 6-32 screw 2 3/4-inches long with washer and wing nut holds the search loop in position.

Put in the batteries and turn on the power to both units. Position the receiver's 4 1/2-inch loop as shown in Fig. 5-7. Watch the meter and turn up the gain until the meter reads 3/4 scale. While watching the meter, slowly move L3 back and forth and around the folded-loop area of L2 until the balance point is found. The balance condition should occur in about the same area as in Fig. 5-7. As the meter reading drops, keep raising the gain (R6) until the best possible balance is obtained. When the locator is in proper balance the gain can be set to maximum, allowing the detector to be used in its most sensitive operating mode. After the balance location has been determined you should mount the loop permanently in place with glue or epoxy. Mounting the loop in its proper position can be a bit tricky, but a little patience and TLC here will pay off later.

If the locator doesn't function properly or the sensitivity is too low, check to make sure the receiver loop is tuned to the same frequency as the oscillator. Move the search loop away from any metal and adjust the gain to maximum. If the balancing procedure was done properly, the meter should read zero or near zero. If you have a zero reading, a small metal object will be needed to cause the meter reading to increase to approximately 1/4 scale. Tape the metal object in place on the loop's plywood base to maintain the 1/4-scale reading during the tuning process.

Disconnect one end of capacitor C1 in the receiver and substitute a value at least 10% smaller in size. If the meter reading increases, a smaller capacitor is needed to replace C1. Keep trying different values until the correct one is found. A capacitor decade box would be most helpful for finding the correct capacitance. If the meter reading drops, the value of C1 must be increased. Reconnect C1 and add a small capacitor, that is at least 10% of C1's value, in parallel with C1. Keep adding capacitors until the meter reading peaks, indicating a properly tuned circuit. After the correct capacitor value is found solder the new capacitor permanently in place.

If there is some other type of circuit trouble, check the voltage

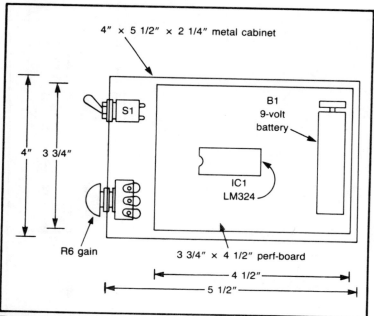

Fig. 5-9. Parts layout of receiver.

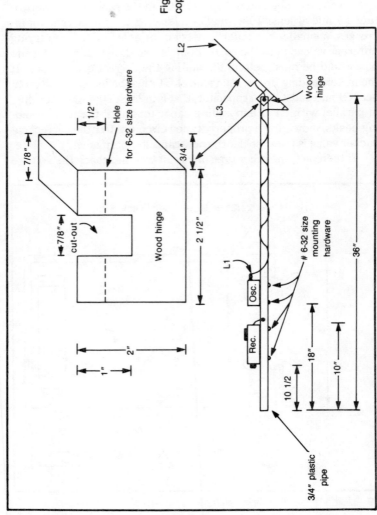

Fig. 5-10. Side-view of the coplanar locator.

Table 5-3. Voltage Readings for Coplanar Locator.

Circuit Location	Voltage
IC Pin #(1)	4.6 volts
IC Pin #(2)	4.6 volts
IC Pin #(3)	4.6 volts
IC Pin #(4)	9.0 volts
IC Pin #(5)	4.6 volts
IC Pin #(6)	4.6 volts
IC Pin #(7)	4.6 volts
IC Pin #(8)	4.6 volts
IC Pin #(9)	4.6 volts
IC Pin #(10)	4.6 volts
IC Pin #(11)	0 volts
IC Pin #(12)	0 volts
IC Pin #(13)	0 volts
IC Pin #(14)	0 volts
Emitter of Q1	− 3.75 volts

readings in Table 5-3. All voltages are taken with a 20,000 ohms-per-volt meter, and IC readings are taken with the receiver on and the oscillator off. The voltage at the emitter of Q1 is taken with the oscillator on, and the positive meter lead connected to the positive battery terminal. If the tuning, and voltage readings are correct, recheck all solder joints and point-to-point wiring to make sure they match the schematics.

EXPERIMENTING WITH THE COPLANAR SEARCH LOOP

Almost any size search loop can be built for the VLF coplanar locator by following a few simple design steps for the coil size and number of turns. The ratio of the transmitter and receiver loop, for all practical purposes, is 2.4 : 1. Just divide the transmitter coil diameter by 2.4 and you will obtain the diameter of both the folded-loop section and also the receiver loop. You can work backwards just as easily. If you know the receiver loop size then the diameter of the transmitter loop can be calculated by multiplying the diameter of the receiver loop by 2.4.

Two search loop sizes can be helpful in locating both small objects and deep buried objects. For very small objects a 6-inch transmitter loop would be ideal. Just divide 6-inches by 2.4 and the diameter needed for the folded-loop and receiver loop is then 2.5 inches. To use the VLF coplanar as a deep-searching locator, a 30-inch transmitter loop would be a good choice. Dividing 30 inches by 2.4 gives a loop size of 12 1/2-inches.

If the receiver coil is wound with approximately 60 feet of number 26 enamel-covered copper wire, it should tune to the oscillator's frequency no matter how large or small the receiver coil is made. Each receiver coil size will need to be tuned to the oscillator frequency, either by using a capacitor decade box or by the trial and error method of adding and subtracting capacitors until the correct value is found.

The number of turns on the secondary winding of L1 may need to be changed to match the impedance of the different sizes of transmitter coils. For the smaller loop sizes try removing a turn or two from the secondary of L1, and for the larger loops try adding a turn or two. A little experimenting may be needed to determine the number of turns on the secondary of L1 that will best match the impedance of each different loop size.

USING THE COPLANAR LOCATOR

The operation of the VLF coplanar locator is very similar to most of the other locators discussed previously. One feature of the coplanar locator that really stands out is the ability of the search loop to pinpoint a buried object with a great deal of accuracy. The target area is small and is centered in the middle of the receiver coil. With a little practice a small metal probe or screwdriver can be used to locate the buried object on the first try.

Unusual Locator Circuits

No matter how many ways you try to classify metal locator circuits, there are a few unusual designs that just don't fit any of the standard categories. This chapter covers several good examples of these interesting, "maverick" locator circuits.

LOADED-LOOP LOCATOR

The first one is a loaded-loop circuit that operates on the principle that a metal object near the search loop robs the oscillator of a small amount of energy. This energy loss can be measured by a meter circuit to indicate the presence of metal.

The parts list and schematic for the loaded-loop circuit are shown in Fig. 6-1 and Table 6-1. The circuit uses a modified Colpitts oscillator to create the basic loaded loop. A variable resistor, R3, controls the oscillator feedback gain. With R3 set to its minimum resistance, the oscillator feedback gain is at its highest, and the oscillator circuit is then insensitive to any loop loading from an outside metal object. If the feedback is reduced to a point where there is just enough gain to sustain oscillation, then the loop becomes very sensitive to outside loading.

To know when the oscillator is adjusted properly you need a method of measuring the rf across the coil. Rf is sampled at the collector of Q1 and feeds a voltage-doubler rectifier circuit. The dc output is prevented from loading the oscillator by an emitter follower, Q2, and the rf output level is monitored by M1.

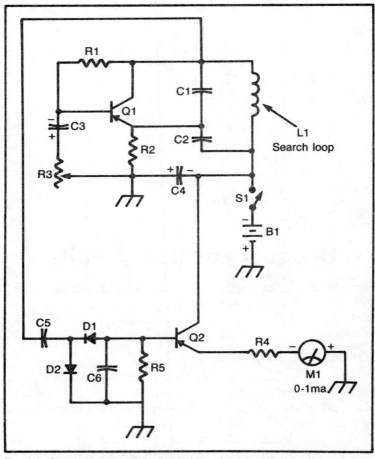

Fig. 6-1. Circuit diagram of the loaded-loop locator.

With R3 set to where the meter reading is one-half scale or less, the locator is ready to seek out hidden or buried metal items. The loaded-loop locator is not as sensitive as some of the other locators, but will find most metal objects within 10 to 12 inches of the search loop. Of course, the size of the metal object is most important, and will determine the depth of detection for any metal locator.

Building the Loaded-Loop Locator

The loaded-loop locator circuit is simple. A good, straight for-

Table 6-1. Parts List for Loaded-Loop Locator.

B1	Transistor-type 9-volt battery
C1, C2	.22 μF/100-volt mylar capacitor
C3	6.8 μF/16-volt electrolytic capacitor
C4	100 μF/16-volt electrolytic capacitor
C5, C6	.1 μF/100-volt mylar capacitor
D1, D2	1N914 silicon signal diode
L1	9-inch diameter search loop with 45 turns of number 26 enameled copper wire
M1	0-1 mA dc meter
Q1, Q2	General purpose pnp silicon transistor, 2N3638A or 2N4249, etc
R1	100 kilohm 1/4-watt 5 or 10% carbon resistor
R2	470 kilohm 1/4-watt 5 or 10% carbon resistor
R3	20 kilohm linear taper pot
R4	1 kilohm 1/4-watt 5 or 10% carbon resistor
R5	470 ohm 1/4-watt 5 or 10% carbon resistor
S1	SPST toggle or slide switch
Miscellaneous	See text

ward parts layout will do just fine. In fact, any of the similar, previous oscillator layout plans can be followed. In any case, since the oscillator frequency is around 16 kHz, no special wiring precautions need to be taken.

The search loop is wound with 45 turns of number 26 enameled copper wire on a 9-inch diameter wood or fiber form. Because of the low operating frequency of the loaded loop oscillator, a Faraday shield is not necessary. The loaded-loop locator is as simple as any to use, and all that's really necessary is to practice using it until operating it becomes second nature to you.

FERROUS MOTION LOCATOR

Figure 6-2 shows a schematic of a most unusual metal detector which responds only to ferrous objects (see Table 6-2 for the parts list). Furthermore, the metal object must be in motion near the pick up loop to be seen by the locator circuit. There are several advantages to using this type of circuit. Any metal item that remains stationary will not be detected—only ferrous metal items that actually move by the pickup loop will cause an output on the meter and LED. A large metal object the size of an automobile can be detected several feet from the pickup loop and could be used with an electronic counter to total the number of vehicles passing a given point. The pickup loop coil can be placed near a conveyor belt to indicate when a ferrous item passes by, or the output can

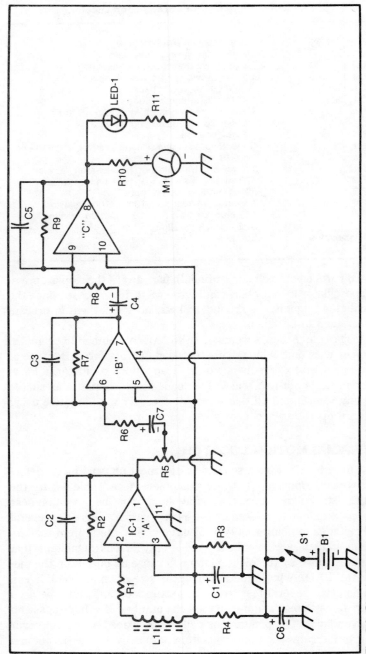

Fig. 6-2. Circuit diagram of the ferrous motion locator.

94

Table 6-2. Parts List for Ferrous Motion Locator.

B1	9-volt transistor battery
C1, C6	330 μF/16-volt electrolytic capacitor
C2, C3, C5	.1 μF/100-volt mylar capacitor
C4, C7	4.7 μF/16-volt electrolytic capacitor
LED 1	Red LED (any general purpose LED will do)
M1	0-1 mA dc meter
IC 1	LM324 quad op amp IC
R1	2.2 kilohm 1/4-watt 5 or 10% carbon resistor
R2, R7	470 kilohm 1/4-watt 5 or 10% carbon resistor
R3, R4	4.7 kilohm 1/4-watt 5 or 10% carbon resistor
R5	10 kilohm linear taper pot
R6, R8, R11	1 kilohm 1/4-watt 5 or 10% carbon resistor
R9	100 kilohm 1/4-watt 5 or 10% carbon resistor
R10	10 kilohm 1/4-watt 5 or 10% carbon resistor
S1	SPST toggle or slide switch
Miscellaneous	See text

be used to operate a relay to stop the conveyor or to flash a warning light.

A much larger open-air pickup loop can be wound and used in place of the small ferrite rod loop. A loop large enough for saw logs to pass through could be used to detect nails, wire, and other ferrous metal objects that could damage an expensive saw blade. A loop large enough to encircle a doorway could be used as a weapon detector. These suggestions are only a few examples of the many ways in which the basic circuit can be applied.

Circuit Operation

Any magnetic field disturbance near L1 is fed to a high gain IC op amp circuit, IC-1A. The signal is boosted over 200 times with the first amplifier, and its output is connected to the circuit gain control, R5.

Op amp B amplifies the signal over 400 times and the output is connected to the input of op amp C where the signal is increased even more by a gain factor of 100 times. The theoretical gain of the three amplifier stages is a staggering 8,000,000 times.

If the op amps were connected in a standard audio amplifier configuration, the circuit would go into oscillation due to the high gain and would not be useful, but with the addition of capacitors C2, C3, and C5 the frequency response of the three-stage amplifier is reduced to near dc. The ultra-low frequency response also helps reduce the possibility of interference from the 60 Hz line frequency.

When a ferrous object passes by the ferrite pickup loop a small magnetic signal is detected and boosted by the three-stage amplifier as a low frequency pulse. The 0-1 milliammeter reads about one-half scale without any input signal, but can drop nearly to zero or rise to full scale as an object passes the pickup loop. Also, the LED will flicker off and on as the object passes the loop. Power for the circuit is supplied by a 9-volt transistor battery.

Building the Motion Detector

The electronics can be housed in any metal cabinet of sufficient size, and the components mounted on perf-board with push-in terminals.

Follow the circuit diagram in Fig. 6-2 to wire the circuit. If the pickup is to be located a distance from the main circuit, use a shielded lead in connecting it to the circuit.

Place two rubber grommets on the ferrite rod, as shown in Fig. 6-3, spaced 2 1/4 inches apart. Approximately 350 feet of number 37 enameled copper wire is wound on the ferrite rod between the two grommets. This is about 3000 turns, but the actual number isn't very important. A coil would with only one half the number of turns, or one with twice the turns would work out all right, so don't waste too much time trying to determine the actual number of turns when winding the pickup coil.

A fast way to wind the coil is to place the ferrite rod material in the chuck of a variable-speed hand drill and mount the wire spool where it can turn freely. Start the drill at slow speed condition and

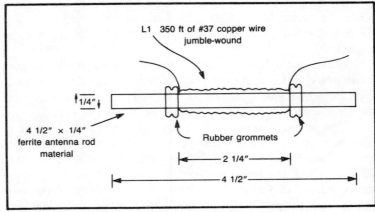

Fig. 6-3. Pickup coil for ferrous motion locator.

slowly increase the speed while winding the coil. Even though the coil is jumble-wound try to keep it neat and as evenly spaced on the rod as possible.

Checking Out and Using the Motion Detector

Any twisted two-wire cable can be used to connect the pickup to the circuit, but if a long run is needed use a shielded two-wire cable and connect the shield to the battery negative of the circuit.

Place the pickup in a fixed position and turn S1 on. The meter should read about one-half scale after the circuit becomes stable. The stabilizing time should be no more than a few seconds for the three amplifiers, even with the gain set to its maximum. A good test for the motion detector is to take a small, permanent magnet and wave it back and forth in front of either end of the pickup loop. The meter and LED should respond in a "bang-bang" action. Keep moving the magnet back and forth while slowly moving away from the pickup. The motion detector should still see the magnet's field three or more feet away from the pickup coil. All other ferrous metals will cause a similar effect, but any item that is magnetic will produce the greatest effect on the detector.

If the detector fails to operate, check the dc voltage at pin numbers 1, 7, and 8 of the IC. All three pins should read about the same voltage, and will be close to one-half of the battery voltage. If not, recheck all wiring for any errors. Another item to check is the resistance of the pickup coil. The resistance should be between 150 and 250 ohms. The coil is the most likely place for a broken or poorly soldered wire, so check it out carefully. Also, don't let the coil drop or hit a hard surface, as the ferrite rod material will break if mishandled.

PLL LOCATOR

The 567 PLL (phase-locked loop) IC houses a most interesting and versatile circuit that is ideal for a simple but sensitive metal locator. The PLL IC can generate a transmitter signal and function as a receiver and detector at the same time. These two circuit functions make up the basic requirements of both T/R and balanced-inductance locator circuits.

The schematic and parts list for small coin-sized and larger object PLL locator circuits are shown in Fig. 6-4 and Table 6-3. The actual sensitivity to small or large metal objects is determined by the size of the two coils used for L1 and L2 and not by the electronics.

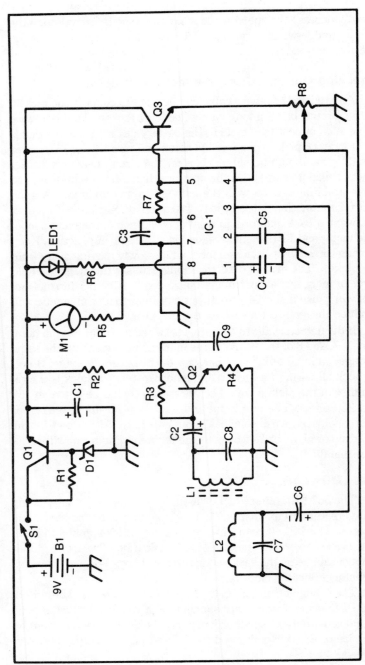

Fig. 6-4. Circuit diagram for PLL locator.

Table 6-3. Parts List for PLL Locator.

B1	9-volt transistor battery
C1	330 μF/16-volt electrolytic capacitor
C2, C4	4.7 μF/16-volt electrolytic capacitor
C3	.0036 μF/100-volt mylar capacitor
C5	.1 μF/100-volt mylar capacitor
C6	10 μF/16-volt electrolytic capacitor
C7	.43 μF/100-volt mylar capacitor (made up by using a .33 μF and a .1 μF capacitor)
C8	.83 μF/100-volt mylar capacitor (made up by using a .5 μF and a .33 μF capacitor)
C9	.05 μF/100-volt mylar capacitor
LED 1	Red LED (any general purpose LED will do)
M1	500 μA dc meter
D1	6-volt zener 1/2-watt 10% diode
Q1, Q2, Q3	2N2222A npn transistor, (any similar general-purpose transistor will do)
IC-1	LM567 phase-locked loop IC
R1, R2, R6	1 kilohm 1/4-watt 5 or 10% carbon resistor
R3	220 kilohm 1/4-watt 5 or 10% carbon resistor
R4	47 ohm 1/4-watt 5 or 10% carbon resistor
R5	10 kilohm 1/4-watt 5 or 10% carbon resistor
R7	33 kilohm 1/4-watt 5 or 10% carbon resistor
R8	500-ohm linear taper pot
S1	SPST toggle or slide switch
Miscellaneous	See text
L1 and L2	Loops L1 and L2 (see text for details)

PLL Circuit Operation

The basic operation of the circuit in Fig. 6-4 is as follows. Transistor Q1 is connected as a dc series voltage regulator, setting the output voltage to about 5.5 volts. IC1 is connected in a basic PLL tone receiver circuit, with the operating frequency set by the values of C3 and R7 to about 6.5 kHz. The PLL will respond with an output at pin number 8 when a 6.5 kHz signal is present at pin number 3.

A 6.5 kHz square wave signal is present at pin number 5 and connects to the base of Q3, an emitter-follower amplifier. The output of the 6.5 kHz signal is fed through R8 to transmitter coil L2. Capacitor C7 tunes the transmitter's loop to the operating frequency of 6.5 kHz. The receiver coil, L1, is positioned on the top of L2 as in Fig. 6-5, in a nearly balanced field location, and is also tuned to 6.5 kHz with C8. When a metal object is positioned over the search loop, the transmitter field is slightly distorted and the receiver loop picks up a portion of the new field created around the metal object. The signal is amplified by Q2 and its output feeds the signal to the input of the PLL, where the meter and LED indicates metal. The most sensitive location on the search loop is just above the end of L1 that connects to C2. A dime-sized object can be detected in open air two to four inches above the end of L1—which is not too bad for such a simple locator.

Building the PLL Locator

The complete electronic circuit can be housed in a metal or plastic cabinet, since shielding at these frequencies is not needed. Perfboard construction with push-in pins is ideal for the construction of this locator. The circuit can be layed out on the perf-board in any suitable scheme as long as good mechanical and electrical construction methods are followed. You can use the basic layout of any of the previous VLF locators as a guide.

The transmitter loop, L2, is wound on a 9-inch form with 45 turns of number 26 enameled copper wire. Make the coil form from wood or fiberboard to match the 9-inch forms illustrated in the earlier units. The receiver coil is wound with 200 turns of the same wire on a ferrite rod 3/8 inches in diameter and 5 1/4 inches in length. Actually, the two coils can be changed in size, shape, and the number turns as long as both are tuned to the PLL oscillator frequency. This PLL locator circuit can be an experimenter's dream, with all the different sizes and shapes of search loops that will work with the basic circuit. A separated-loop, VLF-T/R, search loop arrangement can be built with the same basic PLL circuit. An extra gain amplifier stage can be added between Q2 and the input of the PLL (pin number 3) to give the additional gain required with the separated-loop search head. The search loop can hunt for very small metal objects at shallow depths, or a large search loop can be used for deep searching.

The search loop can be mounted on a wooden carrying stick or handle, but any suitable method will do. The actual arrangement will depend on the type and size of search loop, and as long as the two coils are mechanically stable any scheme should work.

Tuning the PLL

Separate both coils by at least one foot and apply power to the circuit. If an oscilloscope is available, connect the scope input across L2 and turn R8 for a maximum signal to the loop. Connect a capacitor decade box across L2 and select the capacitor value that produces the maximum peak-to-peak voltage across the loop coil. Solder the correct value in place and recheck the tuning.

Connect the scope input to the collector of Q2 and move L1 close enough to L2 to produce a 100 mV signal at the collector. Hook the capacitor decade box across L1 and select the capacitor value that produces the maximum signal at the collector of Q2. Solder the capacitor value in place and recheck the tuning.

Move L2 away from any metal objects and place L1 on the middle of L2 as in Fig. 6-5. While watching the meter and LED display slowly move L1 back and forth until you obtain a meter reading of one-half scale or less. Take a small coin or metal object and move it over the end of L1 which is connected to C2 to determine the most sensitive area. Once the best location of L1 is found, glue or epoxy the coil in place on the transmitter form. The transmitter output control, R8, can be used as a fine sensitivity control.

BALANCED-AMPLIFIER LOCATOR

If the input of an amplifier is coupled to it's own output, the amplifier will oscillate spontaneously, if the signals are in phase.

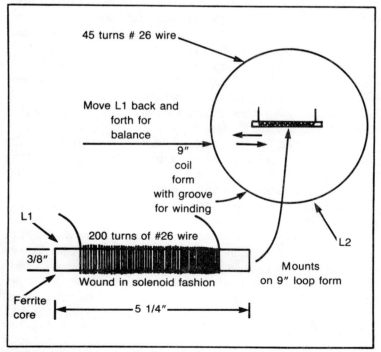

Fig. 6-5. Diagram of search loop for PLL locator.

The only way to stop the oscillation is to interrupt the feedback path or to remove power from the amplifier circuit. In most applications this type of feedback would be a sign of trouble and would require correction before the amplifier could be used.

However, this *feedback oscillator* principle is put to good use in the metal locator circuit in Fig. 6-6 (see Table 6-4 for the parts list). The two coils in the search loop are exactly like the ones used in the PLL locator loop in Fig. 6-5, but almost any other size or shape will work as long as they are properly tuned and each correctly located.

Operation of the Locator Circuit

Op amps A and B of IC 1 form a conventional two-stage audio amplifier circuit. The maximum gain is 10,000 with R10 turned full up. Op amp C is connected as a buffer stage and feeds the rectifier and meter circuit. Any audio signal of sufficient level at the output of amplifier B is rectified by the voltage doubler circuit D1 and D2, with the results indicated on M1.

With the two coils completely balanced in the position shown in Fig. 6-5, there is no signal and the amplifier "sees" only normal circuit noise. However, this is nullified by capacitor C10, which adds a small amount of negative feedback to the isolation amplifier and so prevents any circuit noise from showing up on the meter. If L1 is moved a small amount, circuit noise will be coupled from L2 into the field area of L1 and oscillation will occur instantly. The oscillation frequency is determined by the tuning of the coils, and in this case will be about 6 kHz.

With the coils moved back into balance, however, the circuit will function as a sensitive metal locator. The noise generated by the amplifier circuit is fed to the large loop coil L2, but since the two coils are inductively balanced, no noise is coupled to L1 and M1 reads zero or near zero. When a metal object is moved into the noise field of L2, balance is disturbed, producing a small amount of the noise signal which is sent to L1. The noise is amplified many times, turning the amplifier into a tuned oscillator circuit. As the metal is moved away from the coil field the oscillation stops and the circuit returns to its normal, balanced condition. Whether or not the circuit will oscillate depends on the amplitude of the input signal, which depends on the size and location of the metal object in relation to the search loop. At low gain the locator will respond only to larger metal items; at maximum gain a small coin-sized object can be detected in open air 3 to 5-inches from the search loop.

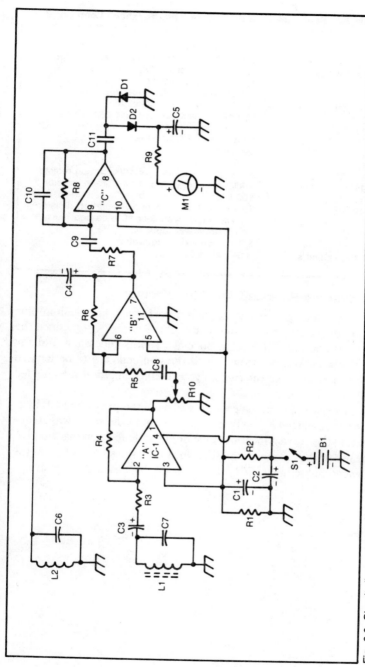

Fig. 6-6. Circuit diagram of balanced-amplifier/balanced-loop locator.

103

Table 6-4. Parts List for Balanced Amplifier/Balanced Loop Locator.

B1	9-volt transistor battery
C1, C2	330 µF/16-volt electrolytic capacitor
C3, C4, C5	4.7 µF/16 volt electrolytic capacitor
C6	.37 µF/100-volt mylar capacitor
C7, C11	.27 µF/100-volt mylar capacitor
C8, C9	.1 µF/100-volt mylar capacitor
C10	680 pF/100-volt ceramic disc capacitor
D1, D2	1N914 silicon diode
L1, L2	Loops L1 and L2 (see text)
M1	0-1 µA dc meter
IC 1	LM324 quad op amp IC
R1, R2	4.7 kilohm 1/4-watt 5 or 10% carbon resistor
R3, R5	1 kilohm 1/4-watt 5 or 10% carbon resistor
R4, R6	100 kilohm 1/4-watt 5 or 10% carbon resistor
R7, R8	10 kilohm 1/4-watt 5 or 10% carbon resistor
R9	2.2 kilohm 1/4-watt 5 or 10% carbon resistor
R10	10 kilohm linear taper pot
S1	SPST toggle or slide switch
Miscellaneous	See text for details

Building the Balanced Amplifier Locator

A metal or plastic cabinet and a perf-board and push-in pins should be used to mount and house all the parts. Again, since the circuit is operating at VLF no special layout is necessary, and any good scheme will do. However, if the two coils are to be located a long distance from the circuit, you should run twin-lead shielded cable to connect them.

The two coils can be constructed as in the previous PLL locator and are placed in the balanced position in the same way. Once you find the correct location, glue or epoxy loop L1 in place on L2. The use of the balanced-amplifier locator is the same as that of the balanced-inductor or T/R-VLF locators discussed previously.

Treasure Hunting

The words "treasure hunting" mean different things to different people. For some people, turning up a few old coins during an afternoon of leisurely coin-shooting is a successful treasure hunt. For a true professional, it may mean months or even years of painstaking research and planning, culminating in the discovery of a large or valuable lost treasure. Regardless of what kind of treasure-hunter you are, this chapter will help you choose the detector that is right for your needs. No matter what the treasure or the type of location, even the best possible locator will fail to perform unless you are experienced in its operation. Complete familiarity with the selected metal locator is a must, and several hours of field, or backyard use is an absolute necessity if the instrument is to be of value in a real hunt. If the practicing is put off until the actual hunt, then you may not do any better than the novice deer hunter who waits until he's in the woods to try his new rifle. Both hunts are apt to fail.

CHOOSING THE BEST LOCATOR FOR THE JOB

The following information should be helpful in selecting the best locator for a specific application. The depth of detection for the frequency-shift type of metal locator (BFO and crystal filter locators) is determined by both the size of the search loop and the size of the metal object. As a general rule of thumb, a metal object 1/10 the size of the diameter of the search loop will produce a de-

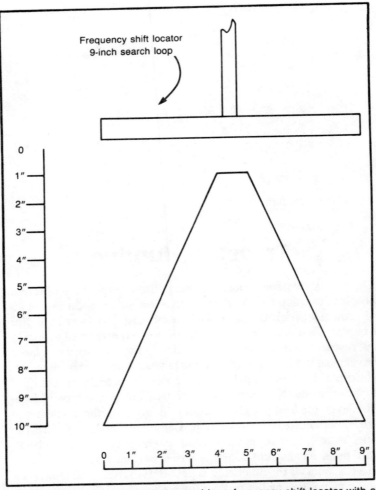

Fig. 7-1. Depth and object size detected by a frequency shift locator with a 9 inch loop.

tectable output. A frequency-shift locator with a nine-inch diameter search loop should be able to detect a metal object the size of a U.S. quarter, but in most cases will detect a dime-sized object only to a depth of 1 to 3 inches. The actual depth of detection with the frequency-shift locator will depend greatly on the moisture and mineral content of the soil and the ability of the operator to "read" the locator output. Also, as a general rule, the depth of penetration is about equal to the diameter of the search loop.

Figure 7-1 shows the depth and object size for a typical frequency-shift metal locator with a 9-inch search loop. A metal object 9 inches in diameter at a distance of 10 inches will produce a frequency shift of about the same amount as an object .9 inches in diameter at a distance of 1 inch from the search loop.

A metal locator with a 9-inch search loop is a good choice for all-around coin or treasure hunting, but a smaller search loop will respond better if only small objects are hunted at shallow depths. Figure 7-2 shows the depth and object sizes to which a frequency-shift locator with a 5-inch search loop will respond. In theory, the 5-inch search loop will detect a 1/2-inch diameter object at a distance of one inch and a 5-inch diameter object at a distance of 5

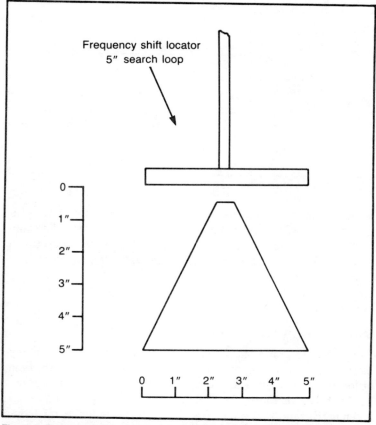

Fig. 7-2. Depth and object size detected by a frequency shift locator with a 5 inch loop.

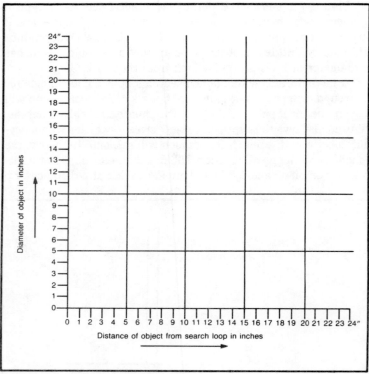

Fig. 7-3. Making a depth vs. size graph.

inches. In actual use the crystal filter type of frequency-shift metal locator will perform much better than Fig. 7-2 indicates. The only way to really ascertain the actual sensitivity and detection range of a given locator would be to make a special depth vs. object size chart similar to the ones in Figs. 7-1 and 7-2. In actual practice making a chart is not very difficult and will help to obtain maximum performance from any locator.

MAKING A DEPTH VS. SIZE CHART

The basic graph show in Fig. 7-3 will work with almost any metal locator, no matter what the size of the search loop or the type of locator. The figures for object size and distance on the graph need only be altered to reflect the size of the locator. The inch values given in the sample graph in Fig. 7-3 are accurate for search loops 9 to 12 inches in diameter. The inch values must be divided in half for coin-shooter locators, making each side 0 to 12 inches.

Take the selected metal locator and move the search loop completely away from any metal object; even a small-loop coin-shooter should be at least three feet away from any sizeable metal object when making the readings for the chart. Several different sizes of round metal material will be needed for making the test. Different size coins, jar lids, can lids, or other metal objects can also be used.

Start out with the smallest round metal object that the locator can detect and move it toward the center of the search loop (bottom), keeping it parallel to the bottom of the loop. Keep moving the metal object back and forth toward the center of the search loop until there is a 15 to 20% increase or change in the locator signal. If a meter indicator is used, mark the chart at the distance where the meter reading changes by 15 to 20%. If an audio tone is your only indication you should repeat the measurements several times.

If a 1-inch diameter metal object causes the meter or tone to change by 15 to 20% at a distance of four inches; mark the chart where the 1-inch diameter vertical line crosses the four- inch horizontal line. Each of the different sizes of objects is tested in the same way and the graph marked to indicate the distance and object size. After all tests have been made connect the dots with a line to complete the graph.

However, just because the test graph indicates a dime-sized object can be detected at a distance of 3 inches this does not necessarily mean that the same level of sensitivity can be expected for all underground locations. After the graph has been completed, a comparison can be made between the open-air tests and underground testing in the outdoor cat box. Don't be surprised if an increase in sensitivity is indicated in the underground tests, because the longer a metal object remains underground the more it tends to mineralize the soil around the outer surface of the buried object. This condition offers the locator a larger target and increases the detection range.

TOOLS

The following three tools will take care of 99% of your needs when coin-shooting.

1. Blunt point ice pick.
2. Standard screwdriver.
3. Hunting knife.

COIN-SHOOTING TECHNIQUES

Any of the small search-loop locators covered in the previous chapters and designed for locating small objects can be used for coin-shooting. Take the selected locator outdoors, or if used indoors, move away from any metal object and air test with different types and sizes of coins. Mentally note the changes in the locator's audio or metered output with each of the objects tested. This information will be helpful later for determining whether to dig for the object or to pass it by. Most of the coin-shooter locators can tell the difference between ferrous and nonferrous metals, and if only coins are hunted the ferrous outputs can be ignored. This feature can probably reduce the digging by a factor of one half, and that is a lot of dirt on a full-day hunt.

BUILDING A "CAT BOX"

If a real outdoor test area is desired, then build a cat box similar to the one in Fig. 7-4. The cost of building the box will depend on the availability of the materials in your area. If a supply of used railroad ties is available, obtain eight ties and place four in a square on level ground, and stack the remaining four on top of the others.

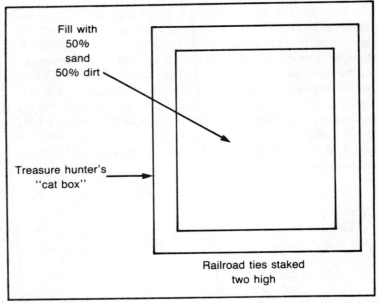

Fill with
50%
sand
50% dirt

Treasure hunter's
"cat box"

Railroad ties staked
two high

Fig. 7-4. "Cat Box."

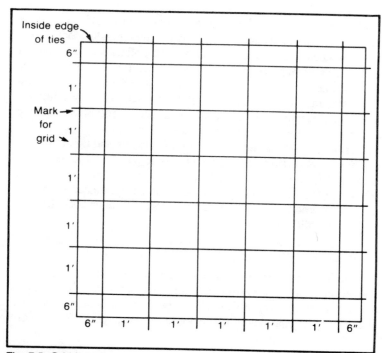

Fig. 7-5. Grid layout.

This double stack arrangement should give ample depth for most buried test items.

Fill the square with a mixture of dirt and sand. A fifty-fifty mix works best, but 100% dirt can be used if obtaining sand is a problem. It would be best to use soil that has a low mineral content.

Mark the wood form, as in Fig. 7-5 and make similar grid on paper, so a grid layout can be used to keep track of the buried items. Each time an item is buried mark the location on the grid sheet. Use a number or letter to designate each of the buried items and record the type and size of the metal object and the depth that it is buried. Use a separate sheet of paper for recording this information.

Small coins can be buried as close as one foot apart, without any problem, but larger items should be separated by a greater distance.

Select a few coins of different denominations, such as copper pennies, clad dimes, and a few nickels. Bury a penny at depths of 1 inch, 2 inches, and 3 inches. Do the same with three nickels and three clad dimes. The 48-cent investment will pay off many times over in the experience gained with each of the metal locators tested. Take two clad quarters and two clad half-dollars and bury one of each at a depth of three inches and six inches. Place all buried coins with the flat side up or parallel to the ground surface. A few selected ferrous metal objects can be included in the cat box. Try using ferrous objects of about the same size as the coins and bury each one at the same depth so comparison tests can be made.

Unfortunately, many lost coins don't always end up in the best position for detection, and only a lot of practice and experience will help you locate these coins. Also, the moisture content of the soil can affect the operation of most electronic metal locators, and it would be a good policy to test out the locator under both dry and wet soil conditions.

WHERE TO COIN-SHOOT

Once you've mastered the operation of your particular locator, you will need a good location in which to look for lost coins. Many of the following locations listed can be found in almost any area of the country, and would make a good starting place for a treasure hunt of lost coins:

- Old yards.
- Public parks.
- Schoolyards.
- Local fairgrounds.
- Old homesteads.
- Ghost towns.
- Sidewalks.
- Around old trees.
- In and around old pathways.
- Outdoor water fountains.
- Outdoor restrooms.
- Children swings.
- Beach areas.
- Church yards.

Any location where people gather year-out is a good place to search for lost coins and jewelry, but no matter where you go to look for coins or treasure always obtain permission from the land owner, or whoever is in charge before digging any holes.

Since almost all good treasure-hunting areas are owned by someone else, as a prospective treasure hunter you should be sure to request the owner's permission to hunt first, or you may find yourself *persona non grata* in the future.

HOW TO SEARCH AN AREA

The beginning coin-shooter often makes the mistake of jumping from one location to another without really covering an area completely. Too often the best finds are missed. A good workable method is to follow a grid pattern similar to the one used in the cat box. Mentally lay out the search area and search over the ground in a straight path. Start the second grid line parallel to the first, but move over by one to two feet depending on the side-to-side area covered in the first search pattern. Continue in the same manner until the desired area is completely covered. Very little will be overlooked using this search method.

It is most important to move the search loop properly over the target area when coin-shooting, so as not to miss a good find. If the search loop is not keep parallel to the ground while scanning a target area, there's a good chance that some of the buried items will be missed or seen as ground effects. Also, it is very important to keep the search loop the same distance from the ground while moving over the target area as in Fig. 7-6. Swinging the search loop back and forth over the target area causes the capacitive effect between the search loop and ground to vary. This effect reduces the sensitivity of the locator and can result in an erratic or constantly changing output.

When the locator gives a good signal, pinpoint the area and take the blunt ice pick and carefully probe the ground until the buried object is found. Slide the screwdriver blade into the ground under the object and then work the object up through the ground. If the grass is too thick take the hunting knife and cut a small opening in the sod above the buried object. After removing the object, carefully replace the dirt in the hole and push the grass back in place. With a little practice a perfect cover-up can be made.

Most coins are found within six inches of the ground surface, and many recently lost coins actually remain on the surface covered only by grass. Always look thoroughly before making a hole since in my experience at least 10% of the coins are on or near the surface.

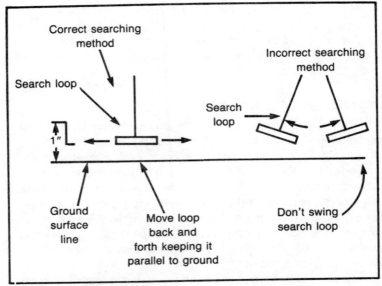

Fig. 7-6. Searching methods.

HUNTING FOR DEEP, BURIED TREASURE

The two-box T/R locator is the best choice for seeking out deep buried treasures or mineral deposits. None of the previously listed locations for coin-shooting are of much value to the deep-searching treasure hunter. A good lead to a buried object is a must if the two-box T/R locator is to be used, as the smaller buried items can not be detected with this type of metal locator. This feature can really be a plus when searching in a "trashy" area where a large number of small metal items are buried.

Once the general area is determined, lay out the area to be searched in the same grid pattern used with the coin-shooter locator. Each search run can be separated by a greater distance, up to three feet between grids, for objects not buried too deeply, and closer if the size and depth of the object are unknown. The amount of time and effort put forth in searching with any locator will depend on what the hidden or buried prize might be.

In any case, the more you use your locator in seeking out buried objects the better you will become at the art of treasure hunting. Experience is still your best teacher.

Index

Other Bestsellers From TAB

Other Bestsellers From TAB

2/96 10X